THIRD WORLD POLITICAL ECOLOGY

Why are the Third World's environmental problems getting worse even though they are receiving unprecedented attention from policy makers and the media? Can these problems ever be solved through incremental policy changes or is their magnitude an indication of the need for a more radical approach? What are the political and economic obstacles to the resolution of the Third World's environmental problems and how might they be overcome? These questions illustrate the need for an approach that integrates political, economic and ecological issues as the basis for an effective response to contemporary environmental problems.

Third World Political Ecology provides an introduction to an exciting new research field that aims to develop an integrated understanding of the political economy of environmental change in the Third World. The authors review the historical development of the field, explain what is distinctive about Third World political ecology, and suggest areas for the future development of the field. The idea of a 'politicised environment' is elaborated so as to clarify the essentially politicised condition of environmental change today. The authors explore in detail the role of various actors – states, multilateral institutions, businesses, environmental non-governmental organisations, poverty-stricken farmers, shifting cultivators and other 'grassroots' actors – in the development of the Third World's politicised environment. They argue that only such an understanding can help to clarify, let alone solve, the environmental problems that plague parts of Asia, Africa and Latin America.

Third World Political Ecology is the first major attempt to explain the development and characteristics of this important new research field. Drawing on examples from throughout the Third World, the book will be of interest to all those who wish to understand the political and economic bases of the Third World's current environmental predicament.

Raymond L. Bryant is a Lecturer in Geography at King's College, London, and **Sinéad Bailey** is a PhD Candidate at King's College, London.

THIRD WORLD
POLITICAL ECOLOGY

Raymond L. Bryant and Sinéad Bailey

London and New York

First published 1997
by Routledge
2 Park Square, Milton Park, Abingdon, Oxon, OX14 4RN

Simultaneously published in the USA and Canada
by Routledge
270 Madison Ave, New York NY 10016

Reprinted 2000

Transferred to Digital Printing 2005

Routledge is an imprint of the Taylor & Francis Group

© 1997 Raymond L. Bryant and Sinéad Bailey

Typeset in Garamond by Routledge

British Library Cataloguing in Publication Data
A catalogue record for this book is available from the British
Library

Library of Congress Cataloguing in Publication Data
Bryant, Raymond L., 1961– .
Third World political ecology / Raymond L. Bryant and Sinéad
Bailey.
p. cm.
Includes bibliographical references and index.
1. Ecology—Political aspects—Developing countries.
2. Environmentalism—Developing countries. 3. Green
movement—Developing countries. I. Bailey, Sinéad,
1972– . II. Title.
JA75.8.B79 1996
304.2—dc20
96–43170
CIP

ISBN 0–415–12743–2 (hbk)
ISBN 0–415–12744–0 (pbk)

This book is dedicated to Sara and Kent

CONTENTS

ILLUSTRATIONS

FIGURES

TABLES

PREFACE

This book seeks to provide an introduction to the rapidly growing research field of Third World political ecology. This field originated in the 1970s, but its real expansion occurred in the 1980s and 1990s. Today, political ecology is a leading source of innovative research on Third World environmental issues. Surprisingly, there is no single book that pulls together the key analytical themes of the field. Rather, the student of Third World political ecology has been forced up until now to seek out a variety of books that address aspects of the subject only (many are now dated as well). For example, Redclift (1987) and Adams (1990) are key works on 'sustainable' or 'green' development; Blaikie (1985) and Blaikie and Brookfield (1987) provide classic accounts on soil erosion and land degradation; *The Ecologist* (1993) and Sachs (1993) tackle questions pertaining to the 'tragedy of enclosure' and 'global environment management'; and most recently Peet and Watts (1996a) link political ecology to discourse theory. The result is a research field whose general 'texts' fail to provide an adequate overview of the field as a whole.

This book seeks to respond to that glaring and important lacuna in Third World political ecology. It does so primarily in three ways. First, the book provides an overview of the historical development of the field since the 1970s, explains why the field has developed in the way that it has, and suggests what is ultimately distinctive about it (and how it might develop in the future). Second, it develops a framework for analysis centred on the idea of a 'politicised environment' in order to clarify the key concepts and presuppositions that underlie much empirical work in Third World political ecology. Finally, the book introduces the reader to that empirical work itself by exploring the main types of actors that are involved in Third World environmental change and conflict. Thus, the discussion of states, multilateral institutions, businesses, environmental non-governmental organisations and grassroots actors (e.g. poor farmers, shifting cultivators, pastoralists) is enlivened with examples drawn from the rich work of political ecologists on Africa, Asia and Latin America. A 'guide to further reading' highlights key books and articles in Third World political ecology for those new to the subject-matter, while a full bibliography provides an extensive list of references for those wishing to explore selected topics or themes in greater depth.

This book has evolved greatly since its conception in 1992. The starting point for the project was a review of the field undertaken by one of us (Bryant, 1992), but formative influences include many of the key works noted above. Blaikie's (1985) *The Political Economy of Soil Erosion in Developing Countries* was especially important in that it approached what was hitherto seen as a 'technical problem' from a new perspective based in political economy. We thus owe to Piers Blaikie, as well as to the other 'pioneers' in Third World political ecology, a special debt of gratitude for having the foresight to pursue political–ecological questions in the first place. The writing of this book has also benefited from the intellectual insights and practical support of a wide range of people over the years who have helped one or both of us. These include Philip Stott, Robert Taylor, Jonathan Rigg, Tim Forsyth, Phil Hirsch, Larry Lohmann, Eric Swyngedouw, Gilbert Braganza, Peter Walpole, Philippe Le Billon, Geoff Wilson, Keith Hoggart, Linda Newson, John Thornes, Chris Hamnett, Margaret Byron and Gordon Reynells. We would also like to thank Tristan Palmer and Sarah Lloyd at Routledge for their editorial assistance, as well as the constructive comments of Bill Adams and Michael Watts on an earlier draft of the book.

Raymond Bryant would also like to thank the students who took his third year undergraduate course on 'Third World Political Ecology', as well as the equivalent MSc course, for their helpful comments and considerable enthusiasm. I am also grateful to those who provided feedback during presentations at the Department of Geography, King's College London (1994), the Institute of British Geographers Annual Conference (1995), and the School of Geography, University of Oxford (1996). I would also like to acknowledge the support of the Committee of Vice-Chancellors and Principals of the Universities of the United Kingdom, the School of Humanities at King's College London, the Nuffield Foundation, and the Economic and Social Research Council (UK) for having supported my research during the period in which this book was prepared. My parents as ever have provided continual support for my research. Finally, my thanks to Sara Thorne for her good-natured support, especially during the final stages of the preparation of this book.

Sinéad Bailey would like to further thank all of those who provided constructive comments on papers presented at the Interdisciplinary Research Network on Environment and Society Annual Conference (1995) and the Department of Geography, King's College London (1996). I am also grateful to the Economic and Social Research Council (UK) for research support during the preparation of this book. My family has been very supportive of my research throughout. My husband Kent Gibson deserves special mention for putting up with my strange work hours, and for helping with the proofreading of this book.

INTRODUCTION

This book is about the politics of environmental change in the Third World, but does not claim to cover all aspects of this vast subject. Rather, our main aim is to explain the key concerns that arise in considering this subject from the perspective of political ecology. In an earlier work, one of the present authors suggested that the emergence of 'Third World political ecology' as a new research field in the 1980s was a reflection of the pressing need for 'an analytical approach integrating environmental and political understanding' in a context of intensifying environmental problems in the Third World (Bryant, 1992: 12). The present study explores that research field in more detail to assess the contribution of Third World political ecology to a broader understanding of the causes and implications of environmental change in the Third World today.

There are a various ways in which scholars have applied a political ecology perspective to the study of empirical problems in the Third World. We examine this question in the context of an overview of the development of the field in Chapter 1. However, three things need to be noted here. First, political ecologists doing research in a Third World context have largely eschewed theory in favour of empirical analysis. As Peet and Watts (1993: 239) note, this situation has resulted inevitably in a research field 'grounded less in a coherent theory than in similar areas of inquiry'. In a recent intervention, though, Peet and Watts (1996b) have suggested that a 'more robust' political ecology would result if the field incorporated theoretical insights from poststructuralist philosophy – a point echoed by Escobar (1996) in his associated call for a 'poststructural political ecology'. It is too early to tell yet whether such calls will be heeded by political ecologists – or whether this theoretical project will even bear fruit in terms of a more coherent perspective on the complex issues surrounding Third World environmental change and conflict (see also Conclusion).

Second, the areas of inquiry comprising Third World political ecology are only broadly 'similar' in so far as scholars have adopted different approaches to the subject-matter. Thus, political ecologists have sought to explain Third World environmental change and conflict in terms of key environmental

1

problems (i.e. soil erosion), concepts (i.e. sustainable development), socio-economic characteristics (i.e. class), actors (i.e. the state) and regions (i.e. South-East Asia), or they have used various combinations of these approaches (see Chapter 1). There is of course nothing wrong with such diversity. Indeed, we would argue that it has rendered the field more robust in the face of the rapid permutation of environmental research generally. Yet there is also a potential downside that political ecologists must be wary of – namely, a research field whose development is hindered by conceptual confusion and fragmentation.

Finally, and to the extent that Third World political ecology already constitutes a coherent research field, it does so in part based on an understanding of political ecology as being predicated on the assumptions and ideas of political economy. Yet the latter encompasses a wide variety of interpretations drawn from across the ideological spectrum – from the political right (neo-classical thought) to the political left (neo-Marxism). To the extent that Third World political ecology is based on ideas drawn from political economy, the field is inevitably caught up in debates over political meaning. Blaikie and Brookfield (1987: 17) may thus be correct in suggesting that Third World political ecology is about the combined 'concerns of ecology and a broadly defined political economy', but this definition is only useful in relation to a clear notion of political economy (or, for that matter, of 'ecology': see Zimmerer, 1994). However, it is our contention that the most effective way of addressing this problem is not through a grand theoretical exposition, but rather through a selective engagement with the political-economy literature as, and when, that literature is appropriate to the argument of this book. The aim is thus to build an appreciation of Third World political ecology's possible multiple interpretations of 'political economy', rather than to assert in a dogmatic fashion a single 'correct' interpretation.

It follows from the last point that this book does not aim to elaborate a theory of Third World political ecology, but rather suggests that a useful way in which to clarify the meaning of the research field is through an analysis of the political interests and actions of the various actors that participate in political-ecological conflict in the Third World. The argument, developed in Chapter 1, is that an actor-oriented approach affords a useful means to elaborate the role of 'place' and 'non-place-based' actors (Blaikie, 1985) in such conflict. The goal is to understand the possibilities for action by actors operating within broader political and economic structures, but without falling into the trap of economic reductionism that weakened some early work in the research field.

A RADICAL PERSPECTIVE

An important reason for the distinctiveness of Third World political ecology as a research field in comparison with other environmental research fields is the radical perspective upon which work in this field is typically based. Chapter 1 notes that political ecologists have moved on in the 1990s from the relatively crude neo-Marxism that characterised some work of the late 1970s and early 1980s. These scholars nonetheless continue to orient their work around radical agenda that usually set them apart from other scholars conducting research on environmental problems in the Third World.

Political ecologists appear to agree on two basic points notwithstanding their differing research approaches. First, they agree that the environmental problems facing the Third World are not simply a reflection of policy or market failures (as, for example, the World Bank would have it: see World Bank, 1992), but rather are a manifestation of broader political and economic forces. Those forces are associated with the worldwide spread of capitalism, notably since the nineteenth century (see Chapters 3 and 5). The work of political ecologists has been largely an attempt to describe the spatial and temporal impact of capitalism on Third World peoples and environments. This work has focused on the adverse social and environmental consequences of capitalistic natural-resource extraction: logging, mining, fishing or cash-crop production (e.g. Watts, 1983a; Chapman, 1989; Hurst, 1990; Moody, 1996). Attention is also turning to urban and industrial pollution issues as selected areas in the Third World undergo rapid industrialisation and urbanisation, and as the First World seeks to unload the toxic byproducts of First World industrial activity on unsuspecting or weak Third World states (Weir, 1988; Leitmann *et al.*, 1992; Douglass, 1992; Setchell, 1995; see also Conclusion).

Yet the source of the Third World's environmental woes cannot be equated with the workings of the global capitalist system alone. Political ecologists thus also emphasise the ways in which the state in the Third World may intervene in economic activity to promote environmentally destructive activities (see Chapter 3). In some cases, such intervention goes hand-in-hand with capitalist expansion, but in other instances it can reflect a ruler's interest in political power, national security or personal enrichment.

What is clear, however, is that the sources of the Third World's environmental problems are sufficiently complex and deep-rooted to belie any 'quick fix' technical policy solution. Thus, a second area of agreement among political ecologists is the need for far-reaching changes to local, regional and global political–economic processes (Peet and Watts, 1996a). These changes will not occur without considerable struggle since they necessitate the transformation of a series of highly unequal power relationships upon which the present system is based: First/Third Worlds, rich/poor or rulers/ruled. Political ecologists do not necessarily advocate a type of anarchistic ecological utopia as some seem to suggest (e.g. Lewis, 1992). However, they do tend to argue that the social and environmental 'contradictions' (Redclift, 1987) of the global capitalist system are such as to negate the effects of all attempts to reform the status quo. Radical change is required if a resolution of the Third World's environmental crisis is to be effected.

It is for this reason that political ecologists are highly sceptical of the merits of the concept of sustainable development (and related 'ecological modernisation' theorising: see Weale, 1992). To be sure, some argue that the emphasis of this concept on development that does not prejudice the livelihood prospects of future generations or which perpetuates existing poverty shows the revolutionary potential of sustainable development (Redclift, 1987). Yet the alacrity with which powerful political and economic actors embraced the concept following publication of the World Commission on Environment and Development's (1987) report *Our Common Future* without at the same time committing themselves to real change (and the 1992 Rio Earth Summit changed little in this regard), is indicative of an assimilation of the concept to 'business-as-usual' agenda (Hecht and Cockburn, 1992; Adams, 1993). Political ecologists are thus increasingly scathing of the utility of sustainable development as a basis for radical change, and suggest generally that mainstream 'reformist' approaches have hit an impasse both intellectually and practically in their efforts to address intensifying environmental problems (Bryant, 1991, 1997b; Middleton *et al.*, 1993; Sachs, 1993; Escobar, 1996).

Political ecologists have yet to develop an alternative to the mainstream concept of sustainable development. Indeed, they are much stronger at criticising the status quo than at developing a feasible blueprint for an alternative political economy (e.g. see the various contributions to Peet and Watts, 1996a). This situation is not surprising since it reflects a wider dilemma facing radical perspectives. As Pepper notes, radical scholars

> can be criticised either for not suggesting a coherent and feasible action programme, or, if they do have a programme, for being naive and/or anodyne about what could and should be done. This is partly because liberal-capitalist assumptions about the purpose of life and how to live it have gained such hegemony that any attempt to move towards a society based on alternative assumptions does seem either undesirable or futile.
>
> (Pepper 1993: 235)

To the extent that it is possible to speak of a blueprint for change linked to Third World political ecology, it would undoubtedly encompass movement towards an ideal type in which local-level decision-making by grassroots actors (understood in this book to include poor farmers, shifting cultivators, fishers, hunter-gatherers, and the like) would figure prominently at the expense of the activities of non-place-based, and traditionally powerful actors (used here to denote states, large businesses or multilateral institutions) (cf. Dryzek, 1987; Norgaard, 1994). The latter group of actors would still have a role to play, but it would be a supportive role that would revolve around coordinating the interaction of local-level actors and communities at various scales. Decision-making in such a political system would be animated by considerations of social justice (i.e. equity) and environmental conservation (i.e. sustainable local livelihoods) (Chambers, 1987; Hecht and Cockburn, 1992). Such an ideal-type follows logically from the widespread linking of the actions of non-place-based, and traditionally powerful actors to social injustice and environmental degradation in most political ecology work (including this book). Yet if political ecologists are somewhat vague on the contours of an alternative political economy, they are much clearer on the need to emphasise the role of politics, both in terms of understanding contemporary political–ecological conflict, and in devising a way out of the Third World's growing environmental crisis.

PUTTING POLITICS FIRST

At the heart of research in Third World political ecology is the idea that politics should have analytical pride of place. To some extent, the primacy of politics reflects the origins of the research field as a response to the apolitical nature of most research on Third World environmental problems in the 1970s (see Chapter 1). As Deutsch (1977: 359) noted at the time: 'political processes and institutions are rarely mentioned directly and even more rarely analysed in detail. And yet, the substance of politics ... is inescapably implied in almost every ecosocial problem.'

Yet the importance of the 'political' in political ecology represents much more than simply an attempt to correct the perceived deficiencies of other environmental research fields. It reflects above all a reasoned argument that to understand, as well as to be in a position to solve, the Third World's environmental crisis is to appreciate the ways in which the status quo is an outcome of political interests and struggles. Indeed, it is to acknowledge the very existence of a 'politicised environment' in the Third World (see Chapter 2). Thus, politics and environment are everywhere thoroughly interconnected. As Harvey notes,

> all ecological projects (and arguments) are simultaneously political-economic projects (and arguments) and vice versa. Ecological arguments are never socially neutral any more than socio-political arguments are ecologically neutral. Looking more closely at the way ecology and politics interrelate then becomes imperative if we are to get a better handle on how to approach environmental/ecological questions.
>
> (Harvey 1993: 25)

Yet the relationship between politics and ecology is not an equal one. In effect, the role of politics in shaping ecology is much greater today than in the past as a result of rapid social and technological changes that render problematic the idea of a 'natural' environment (McKibbens, 1989; Blaikie, 1995b), and may soon bring under human control the operation of ecological processes themselves. In contrast, the purchase of ecological processes on human affairs has long been in decline, and appears set to continue to decline in the future as 'produced nature' becomes the norm (Castree, 1995; Escobar, 1996; and see Chapter 2).

It is partly for this reason that political ecologists tend to favour consideration of the political over the ecological (this is also linked to the social science background of most political ecologists: see Chapter 1). It is true that political ecologists ought not to ignore advances in the understanding of ecological processes derived from the 'new ecology' since, in doing so, they might miss an important part of the explanation of human–environmental interaction (Zimmerer, 1994; see also Blaikie, 1995b; Forsyth, 1996). Yet greater attention by political ecologists to ecological processes does not alter the need for a basic focus on politics as part of the attempt to understand Third World environmental problems. After all, the development of those problems is linked to political processes. Further, it is only through political means that a solution to those problems will be devised. The need for political analysis is paramount in the measure that the environment is politicised.

Ironically, Moore (1993) has recently criticised much research in Third World political ecology as being insufficiently political in nature (reiterated in a revised piece: see Moore, 1996). As a result of the field's 'structuralist legacy', Moore alleges, there has been too little effort given to exploring the rich 'micro-politics' that condition environmental conflict in the Third World. Further, the interests and actions of the actors involved in such conflict were often portrayed as 'monolithic' in that little effort was made to appreciate the internal complexity or differentiated concerns of the state or other actors (Moore, 1993: 381). Moore's critique holds considerable validity with reference to political-ecology research of the late 1970s to mid-1980s when economic determinism largely prevailed (Bryant, 1992). However, it is a measure of how far the field has moved since that time that the critique appeared increasingly outdated by the mid-1990s. Diverse studies now document political and ecological interactions at the micro-level (e.g. various articles in Neumann and Schroeder, 1995). Further, and as this book suggests, an appreciation of the complex interests and actions of place and non-place-based actors in environmental conflict is now an integral part of Third World political ecology.

Indeed, we argue that, if anything, the research field suffers today from an exaggerated emphasis on local activities at the expense of regional and global developments. As Chapter 1 notes, there has always been an anthropological-style preoccupation with local problems in Third World political ecology even when those issues were linked to structural forces. Thus, although Moore (1993, 1996) is correct in opposing structural determinist explanations of local-level conflicts, he is wrong in suggesting (after Neumann, 1992, and Blaikie and Brookfield, 1987) that the research agenda as a whole need to be centred on local 'land' users (Moore, 1993: 381). Such a perspective can only limit the appeal and utility of political ecology at a time when the Third World's environmental problems are becoming much more heterogeneous in nature. In contrast, the argument of this book is that Third World political ecology must address the politics of all types of environmental change, and at various scales, in so far as they have an impact on the Third World.

THE THIRD WORLD CONTEXT

It could be argued in an era of 'globalisation' characterised notably by the collapse of the Second World that there is no longer a need for a political ecology of the Third World. Is it not time for a 'global political ecology' (Swift, 1993) that does away with regional connotations in a 'world without frontiers'? It is our view that, notwithstanding the political and economic upheavals of recent years, a specifically 'Third World' political ecology is still essential to an appreciation of the environmental crisis besetting the Third World.

To begin with, those upheavals need to be set against a long period of development that is rooted in distant colonial times, but which even today powerfully conditions the way in which human–environmental interaction takes place in the Third World. The manner in which the 'people without history' were incorporated into a European-dominated international economic order in the five hundred years following Columbus's 'discovery' of America in 1492 has been widely documented (Wolf, 1982; Blaut, 1993). The role of the colonies in Asia, Africa and Latin America in this economic order was to provide precious metals, spices, tea, coffee, timber, minerals, cotton, groundnuts, copra and other products for consumption or manufacturing in Europe and North America. However, environmental degradation was often the result as large areas of forest were felled for timber or cleared to make way for export-oriented cash crop production (Rush, 1991; Miller, 1994). In other areas (e.g. West Africa), lands already dedicated to multi-purpose agriculture were converted (often forcibly) to cash-crop monocultures of cotton or groundnuts wanted in Europe and North America, but at the cost of widespread soil degradation in the producing areas (Franke and Chasin, 1980; Watts, 1983a).

Yet the colonial legacy in the Third World is more than one of environmental degradation and economic dependency on natural resource exploitation. Colonial rule also led to political and administrative changes that fundamentally altered the ways in which states went about managing the peoples and environments under their jurisdiction. This issue is explored in Chapter 3, but suffice it to say here that new scientific and administrative techniques widely adopted in the nineteenth century considerably enhanced the ability of the state to control human–environmental interaction within its jursidiction. A colonial legacy of integration in a global capitalist economy, natural resource dependency, environmental degradation and centralised political control have conditioned environmental use and conflict in postcolonial times. That legacy, in turn, continues to distinguish the Third World from elsewhere.

There is thus little to indicate that 'the end of the Third World' (Harris, 1986) is nigh. The remarkable economic success of Asia's Newly Industrialising Countries (the NICs: Singapore, Taiwan, Hong Kong, South Korea) is certainly not to be gainsaid, and such success is spreading to neighbouring Malaysia, Thailand, Indonesia and parts of southern China. However, rapid economic growth in the latter countries has yet to alter materially the conditions of poverty that afflict the vast majority of people in those countries. Further, even in the NICs themselves prosperity is unequally distributed, and in any case, the total population of all four countries is minuscule in relation to the total population of the Third World. As such, it is premature to herald the end of the Third World based on the evidence of (uneven) economic prosperity in a handful of countries in Asia. It is unclear, moreover, whether that success can be replicated on a large scale elsewhere in the Third World (Cline, 1982).

Indeed, it is the fact that extreme poverty is a way of life for the vast majority of people in the Third World (despite fifty years of 'development') that provides perhaps the strongest justification for a specifically Third World political ecology. Although poverty in the First and Second Worlds is not to be denied, the sheer scale of the poverty problem in the Third World stands out. Such poverty, in turn, ensures that environmental conflicts in the Third World (as opposed to the First World where aesthetic concerns often prevail) are predominantly livelihood based. Indeed, a key research objective in Third World political ecology is precisely to explore the connections between poverty and wealth, environmental degradation and the political process (Redclift, 1992; *Ecologist*, 1993; Broad, 1994).

There are thus clear grounds for research in political ecology that is focused on the Third World and which is distinct from research on other areas of the world. Indeed, the history of political-ecology research in general tends to corroborate this view. Thus, research in Third World political ecology has taken place in virtual isolation from theoretical and empirical developments associated with First World political ecology (e.g. Black, 1990; Atkinson, 1991; Roussopoulos, 1993; a notable, if partial exception is Blaikie and Brookfield, 1987). Yet, there have been numerous efforts to link political-ecology research on different parts of the Third World to explore common themes (e.g. Little and Horowitz, 1987; Blaikie and Unwin, 1988; Repetto and Gillis, 1988; Friedmann and Rangan, 1993; Neumann and Schroeder, 1995; Peet and Watts, 1996a). Conversely, there is an astonishing lack of recognition in First World political ecology that a vibrant counterpart research field exists with reference to the Third World (e.g. Atkinson, 1991; Roussopoulos, 1993). The First/Third World divide is thus nowhere more evident than in the division of intellectual labour amongst scholars who conduct political-ecology research. It may be the case that future political and economic developments worldwide will lead to a blurring of this intellectual divide but in the interim there are tangible benefits to be gained from such an approach to the subject-matter.

This book therefore seeks to explore the contribution of Third World political ecology to an understanding of the politics of environmental change in the Third World. The goal is not to review the existing literature in the field, or to provide yet another empirical case study of political and ecological problems. Rather, the objective is to consider the broad concerns that animate this emerging research field. Our approach, which is spelled out in Chapter 1, is to focus on the key types of actors involved in environmental conflict in the Third World, and to use specific examples from Africa, Asia and Latin America to illustrate selectively the general arguments. Before doing so, however, it is necessary to provide an overview of the origins and development of Third World political ecology.

1

AN EMERGING RESEARCH FIELD

For all its 'extraordinary vitality' (Peet and Watts, 1993: 242), the origins and development of Third World political ecology remain poorly understood. Scholars adopting a political ecology perspective have rarely paused to consider the history of the field or its disciplinary location. Rather, the emphasis has been on the empirical application of a broadly defined political economy to the political and ecological problems of the Third World. Yet to understand the significance of research in any field of inquiry is partly to appreciate the development of that field. This chapter presents an overview of the origins and development, disciplinary location and approaches to Third World political ecology.

ORIGINS AND DEVELOPMENT

The field of Third World political ecology originated in the early 1970s at a time when human–environmental interaction was coming under close public and scholarly scrutiny, especially in the First World. A paper by Wolf (1972) is seen as one of the earliest works in what would eventually become Third World political ecology, but no 'classic' work marked the advent of the field. Rather, it developed slowly in light of research developments in other fields, especially cultural ecology and radical development geography. Indeed, until the publication of Watts (1983a), Blaikie (1985), Bunker (1985) and Hecht (1985) the field could scarcely be said to have existed at all.

There are several reasons for the slow development of Third World political ecology in the 1970s. To begin with, the term 'political ecology' had strong negative connotations for many on the political left during that decade as a result of its association with the work of Ehrlich (1968), Hardin (1968), Heilbroner (1974) and Ophuls (1977). These 'eco-doomsayers' predicted that the world faced imminent social and environmental catastrophe due to runaway population growth (Third World) and consumption levels (First World). They argued that an authoritarian global state or 'Leviathan' was needed to enforce 'limits to growth' (Meadows *et al.*, 1972; Ophuls, 1977; see also Chapter 3). Although this school of thought has been usually known under the name 'neo-Malthusianism' (for its anxiety over the population question), its penchant for drastic political prescriptions to solve the world's environmental crisis led to its work also being described from time to time as political ecology. The critique of such 'political ecology' was mounted on the left by Enzensberger (1974) and Harvey (1974), but occurred right across the political spectrum (e.g. Beckerman, 1974). By the late 1970s, the work of this 'political ecology school' had been largely discredited, and attention thereafter turned – via a 'red–green' debate (Weston, 1986; Ryle, 1988) – to the possibility of a convergence between a 'radicalised' political ecology and socialism. The convergence issue has yet to be sorted out in First World political ecology (Atkinson, 1991; Pepper, 1993; Dobson, 1995). However, what is important to this book's discussion of the development of Third World political ecology is the association of Ehrlich and company with the term 'political ecology' in the 1970s. In effect, this association decreased the likelihood during that decade at least that radical environmental scholarship dealing with the Third World would pass under the political ecology label.

And yet, ironically, the work of Ehrlich et al. indirectly influenced the development of a distinctly 'Third World' and radical political ecology. To begin with, work in the field was shaped by the reaction to the simplistic claims about population growth and environmental degradation in the Third World that were at the heart of neo-Malthusianism. The latter was in effect a boon to serious scholarship on Third World environmental change because, however misguided it may have been intellectually, the neo-Malthusian argument nonetheless publicised the importance of such change, thereby prompting questions about causation left unanswered in their work.

That work also prompted a growing interest in the connections between politics and environmental change that is at the core of Third World political ecology today. An important criticism of the neo-Malthusians was that, notwithstanding their association with the term 'political ecology', there was little that was truly 'political' in their analysis (Walker, 1988; Adams, 1990). Ehrlich et al. stood accused of ignoring the political obstacles to, and implications of, the global authoritarian state that they argued was needed to solve the world's environmental crisis. It was only a small step from this critique to a recognition of the need to understand environmental change in the Third World itself as a political process (cf. Cockburn and Ridgeway, 1979; Yapa, 1979).

Neo-Malthusianism also facilitated the emergence of radical development geography, a sub-discipline of geography that has been particularly influential in the development of Third World political ecology. Although radical development geography was part of a larger revision of geography beginning in the late 1960s, it gained momentum in the 1970s partly as a result of its long-running campaign against neo-Malthusianism. For example, work by Buchanan (1973), Darden (1975), Lowe and Worboys

(1978), and Wisner *et al.* (1982) published in the journal *Antipode* attacked diverse aspects of the neo-Malthusian viewpoint, and was part of a broader assault on mainstream environmental research for its neglect of questions derived from political economy (Corbridge, 1986; Adams, 1990).

A sense of what such questions might mean for research was embodied notably in work by radical geographers on 'natural' hazards and disasters. O'Keefe (1975) and Wisner (1976, 1978) initiated a process of inquiry into the interaction of political-economic structures with ecological processes that culminated in alternative research agenda published on the subject of disasters and hazards in the early 1980s (Susman *et al.*, 1983; Watts, 1983b). These agenda were focused on disasters and hazards, but were simultaneously a wider comment about the need for work on the political economy of environmental change in the Third World. As such, they were an influential strand in the development of Third World political ecology, a point acknowledged in key political-ecology texts (e.g. Blaikie, 1985; Blaikie and Brookfield, 1987).

A separate strand in the evolution of Third World political ecology relates to work on environmental topics in anthropology during the 1960s and 1970s. Cultural ecology (or ecological anthropology) sought to explain the links between cultural form and environmental management practices in terms of adaptive behaviour within a closed ecosystem (Bennett, 1976; Hardesty, 1977; Orlove, 1980; Ellen, 1982). However, the emphasis on energy flow modelling and systems analysis resulted in a general unwillingness or inability to see that the local-level cultural and ecological communities being studied formed part of (and were influenced by) a much wider set of political and economic structures (Peet and Watts, 1993; Simmonds, 1993). This work used 'ecology' to emphasise the homeostatic and apolitical nature of human–environmental interaction (Adams, 1990).

However, cultural ecology had become the subject of growing criticism by the early 1980s. Hjort (1982) and Grossman (1984), for example, emphasised the need to couch anthropological insights about human–environmental interaction in the context of an appreciation of the wider political and economic structures that influence activity in any given locality – what Vayda (1983) termed 'progressive contextualization'. The integration of anthropological-style local research with political-economic structural analysis thereafter became a key concern of political ecologists (e.g. Hecht, 1985; Little and Horowitz, 1987; Bassett, 1988).

Political ecologists seeking to integrate place- and non-place-based analysis turned mainly to neo-Marxism in the late 1970s and early 1980s. The latter was heterogeneous in nature and encompassed dependency theory (e.g. Frank, Cardoso, Faletta), world-systems theory (e.g. Wallerstein) and modes of production theory (e.g. Rey, Meillassoux). This work has been reviewed extensively elsewhere (Taylor, 1979; Peet, 1991; Hettne, 1995), but what is important to note here is that neo-Marxism was at its most influential in the social sciences at a time when many political ecologists sought a radical theory to inform their contextual analyses. To be sure, resource depletion had long been a theme in Marxist scholarship, even in a Third World context (Frank, 1967; Caldwell, 1977). However, Redclift (1984: 13) rightly notes that 'Marxist writing about the development process has accorded a secondary role to the natural environment' – a point explored subsequently in debates about ecology and Marxism (O'Connor, 1988; Benton, 1989; Grundemann, 1991; Castree, 1995). Yet, for many political ecologists writing on the Third World in the first half of the 1980s, neo-Marxism offered a means to link local social oppression and environmental degradation to wider political and economic concerns relating to production questions (Blaikie, 1985; Bunker, 1985).

During the first phase of Third World political ecology, which can be said to have run from the late 1970s until the mid-1980s, scholars resorted to neo-Marxism as a way of avoiding the perceived apoliticism of work by many cultural ecologists and neo-Malthusian writers (Table 1.1). Watts (1983a), Blaikie (1985) and Bunker (1985), for example, situated their studies on northern Nigeria, soil erosion, and the Amazon, respectively, in a structural framework informed by neo-Marxist ideas. Although these studies provided rich empirical insights, the emphasis on structure tended at times to downplay the ability of politically or economically weaker grassroots actors, such as small-scale farmers or shifting cultivators, to resist their marginal status. However, the neo-Marxist basis of Third World political ecology at this time was nowhere more evident than in work by Cliffe and Moorsom (1979), Hedlund (1979) and O'Brien (1985) published in the *Review of African Political Economy* which explained local environmental conflicts in terms of class relations and surplus extraction linked to global capitalist production. The role of local politics in mediating resource access and conflict was thereby often largely neglected, and discussion of different actors (i.e. states, businesses, farmers) verged at times on the simplistic (Moore, 1993). The state, for example, was typically seen as being little more than an agent of capital, thereby obscuring both the potential autonomy of this actor vis-à-vis capital, and the diversity of bureaucratic interests that the state often encompasses (see Chapter 3).

Concerns over the influence of deterministic neo-Marxism on the field's development led in the late 1980s to the start of a second phase in Third World political ecology that has drawn on a more eclectic range of theoretical sources. Blaikie and Brookfield (1987), Hecht and Cockburn (1989) and Guha (1989) initiated this process with work on land degradation, the Amazon and India, respectively, but were soon followed by a flood of studies by other scholars (e.g. Peluso, 1992; Neumann, 1992; various articles in Watts and Peet, 1993 [revised and updated as Peet and Watts, 1996a], and in Neumann and Schroeder, 1995) all of which sought to demonstrate a more complex understanding of how power relations mediate human–environmental interaction than was hitherto the case. In doing so, political ecologists

Table 1.1 Phases of Third World political ecology

Phase	Critical of	Theoretical bases	Explanation	Key problem	Major works
1 Late 1970s to mid- 1980s	Neo- Malthusianism & cultural ecology	Neo-Marxism (Frank, Wallerstein, Amin, de Janvry)	Structural. Explain local conflicts/change as outcome of global production process	To specify patterns of surplus extraction & class relations	Watts (1983a), Blaikie (1985), Bunker (1985)
2 Late 1980s to 1990s	'Deterministic' neo-Marxism	Neo- Weberianism (Skocpol, Tilly) & social movements theory (Scott) & household/femin- ist (Guyer & Pe- ters, Agarwal)	Agency. Explain conflict/ change at all scales as outcome of interaction of various actors possessing unequal power capabilities	To specify unequal power relations between actors & motives & interests of various actors	Blaikie & Brook- field (1987), Guha (1989), Peluso (1992), Neumann & Schroeder (1995)

have linked their research to a diversity of theoretical literatures that defy easy classification. Thus, scholars have drawn on neo-Weberian theorising in political sociology (Skocpol, 1985; Mann, 1986) to explore the implications for environmental conflict of the potentially autonomous state (Peluso, 1992; Bryant, 1997a). The potential power of grassroots actors such as poor farmers and shifting cultivators in environmental conflicts has been emphasised with reference to the concepts of avoidance behaviour (Adas, 1981) and everyday resistance (Scott, 1985), as part of an attempt to link political ecology to developments in social movements theorising (Guha, 1989; Peluso, 1992). Scholars influenced by household studies (Guyer and Peters, 1987; Berry, 1989) and ecofeminist writings (Agarwal, 1992; Jackson, 1993) have examined how power relations within the household influence the control of land, natural resources, labour and capital (Carney, 1993; Schroeder, 1993). Finally, and more recently, work has started to draw upon 'poststructuralism' and 'discourse theory' (Said, 1978; Bhabha, 1994; Escobar, 1995) to map the ways in which knowledge and power may interrelate so as to mediate political-ecological outcomes (Fairhead and Leach, 1995; Fortmann, 1995; Jewitt, 1995; Peet and Watts, 1996b).

By the mid-1990s, scholars seemed largely content to explore the politics of the locality as the field moved away from 'political ecology's structural legacy' (Moore, 1993: 381). Yet we would argue that this trend represents an over-reaction to earlier developments. Instead, we suggest that, in an era of 'globalisation', Third World political ecology must develop 'rigorous analyses which link local level production processes and decision making with the larger political economy to explain these different experiences' (Bassett, 1988: 469) – keeping in mind, all the while, the need for contingency and flexibility in explanation (Blaikie and Brookfield, 1987).

The preceding overview has explored briefly the origins and development of Third World political ecology. However, to appreciate fully the development trajectory of the field, and its potential for future growth, it is necessary to consider next how the field relates to contiguous environmental research fields.

MAPPING ENVIRONMENTAL RESEARCH

The development of Third World political ecology is part of a wider process of change that has witnessed the emergence of various environmental research fields in the social sciences and humanities since the 1960s. These fields reflect diverse disciplinary interests and concerns, a detailed discussion of which is beyond the remit of this book. Rather, it is instructive here to map selectively this environmental research so as better to appreciate the disciplinary location and significance of Third World political ecology. The point is not to suggest in doing so that environmental research necessarily falls neatly into one disciplinary category or another. Instead, it is simply an attempt to clarify key traits of Third World political ecology through comparative discussion.

Table 1.2 sets out the key characteristics of selected environmental research fields in relation to Third World political ecology. Four aspects to the information contained in this table merit comment here. First, Table 1.2 illustrates the uneven treatment of issues of Third World environmental change by the different research fields. The prominence of cultural ecology (and ecological anthropology) is to be expected given anthropology's traditional Third World focus (Croll and Parkin, 1992). Human ecology, with long-standing links to geography (e.g. Eyre and Jones, 1967; Whyte, 1986), is also relatively strong in terms of Third World research. More surprising, perhaps, is the relative neglect of Third World issues by the other major research fields. Environmental history, to be sure, provides coverage of the Third World, but its primary focus is still on human–environmental interaction in North America and Europe (Worster, 1988). Environmental economics also tends to privilege First World environmental problems over those of the Third World (Pearce *et al.*, 1989). Environmental management has also been largely concerned with First World problems (O'Riordan, 1995), as has environmental politics (Young, 1992; Garner, 1996).

The virtual absence of environmental politics from Third World research has been especially significant in light of the traditionally apolitical nature of much Third World research. In effect, what little research that has been conducted on environment and politics in the Third World has tended to pass uncritically as political-ecology research. However, this situation is now beginning to change as political scientists turn in growing numbers to Third World environmental topics (e.g. Guimaraes, 1991; Hurrell, 1992).

Table 1.2 Key characteristics of selected environmental research fields (in relation to Third World political ecology)

Field	Geographical focus TW/FW[1]	Main discipline(s)	Date of origin[2]	Major academic journal(s)	Key themes	Affinities with Third World political ecology	Environmental outlook
Cultural ecology (& ecological anthropology)	TW (+ some FW)	anthropology	1960s	yes	Cultural adaptation to environment	strong	ecocentric
Ecological economics	FW	economics	1980s	yes	Integrate economics & ecology (radical)	potentially strong	ecocentric
Environmental economics	FW	economics	1970s	yes	Integrate economics & ecology (mainstream)	weak	technocentric
Environmental history	FW (+ some TW)	history	1970s	yes	History of human–environmental interaction	moderate	ecocentric
Environmental management	FW	geography	1970s	yes	Planned environmental manipulation	weak to moderate	technocentric
Environmental politics	FW	political science	1970s	yes	Green political theory, politics, state, inter-state relations	weak to moderate	technocentric
Environmental sociology	FW	sociology	1980s	no	Greening of social interaction and attitudes	weak	ecocentric
Global ecology	globe	physical sciences & geography	1980s	yes	Global environmental change	weak	technocentric
Human ecology	TW + FW	geography	1960s	yes	Integrated physical & social scientific understanding of human–environmental interaction	moderate	ecocentric
Third World political ecology	TW (+ FWPE)	geography	1980s	no	Political economy of environmental change	—	ecocentric

Notes: (1) Third World = TW; First World = FW; Political Ecology = PE (2) Approximate date for emergence of the field

The question thereby raised relates to the difference between Third World political ecology and environmental politics. We would suggest that the main difference between the two resides in the contrasting theoretical and empirical concerns that arise from their different disciplinary 'homes'. Environmental politics is a research field within political science that applies traditional political questions to environmental matters. It thus examines green political theory, the impact of green issues on the formal political process, the state's role in environmental management, and global environmental politics (Young, 1992; Dobson, 1995; Vogler and Imber, 1996). In contrast, Third World political ecology today resides primarily within geography (see below), and explores the political dimensions of human–environmental interaction. Such an exploration overlaps to some extent with selected concerns in environmental politics, notably with regard to a shared interest in state environmental management practices (see also Conclusion). Nonetheless, political ecology encompasses a wider under-standing of 'politics' than is traditionally found in environmental politics. In effect, the former addresses a diversity of non-state political interests and activities in 'civil society' that tend to be neglected by the latter. A concern with the spatial aspects of human–environmental interaction also tends to set political ecology apart from environmental politics. Above all, political ecology assesses the implications of a 'politicised environment' (elaborated in Chapter 2), while environmental politics only considers the environment in so far as it intrudes on the formal political process.

Second, Table 1.2 suggests that Third World political ecology falls predominantly within geography, but it has as yet to set down firm roots in that discipline. This chapter has already noted the strong influence of radical development geography on Third World political ecology, and the field also resonates well with the sub-disciplines of political geography (Dalby, 1992) and industrial geography (Muldavin, 1996). Not surprisingly, therefore, many political ecologists are trained and/or based in this discipline (e.g. Adams, Bailey, Bassett, Bebbington, Blaikie, Brookfield, Bryant, Carney, Grossman, Hirsch, Jewitt, Jarosz, Muldavin, Rangan, Rigg, Schroeder, Stott, Swyngedouw, Watts, Zimmerer). However, we have also highlighted the influence of anthropology (via cultural ecology) on Third World political ecology. That influence is reflected in the fact that a number of political ecologists have had links to anthropology (e.g. Colchester, Hecht, Horowitz, Little, Moore, Schmink) or the contiguous discipline of sociology (e.g. Bunker, Guha, Peluso, Redclift). Thus, Third World political ecology can be characterised as being a geography-based research field that nonetheless maintains strong links to anthropology and sociology. The field's links to the latter two disciplines have undoubtedly sharpened the empirical insights of Third World political ecology. However, these links may also have delayed recognition of Third World political ecology as an emerging field within geography itself by a geography profession unsure as to whether it is 'bona fide geography'. In this regard, Third World political ecology's disciplinary status is somewhat akin to that of environmental management, human ecology and global ecology – all affiliated to geography to a greater or lesser extent, but with important links to other disciplines (Clarke and Munn, 1987; Steiner and Nauser, 1993; Wilson and Bryant, 1997).

In contrast, the other environmental research fields noted in Table 1.2 have firm disciplinary links with a single key discipline. Cultural ecology (ecological anthropology) has long been recognised as an integral and important part of anthropology (Orlove, 1980). Environmental history and environmental economics, meanwhile, are firmly established as fields within history and economics respectively, while environmental politics and environmental sociology are following suit with reference to political science and sociology (Worster, 1988; Turner, 1995; Buttel, 1996; Garner, 1996).

A further 'measure' of the disciplinary status and location of Third World political ecology relates to the publication outlets used by political ecologists in the dissemination of their work. What stands out here is the absence of a major international English language scholarly journal specifically dedicated to research in the field (in French, there is *Ecologie politique*). Instead, political ecology research is scattered throughout the main geographical journals (e.g. *Political Geography, Economic Geography, Antipode, Geographical Journal, Society and Space, Annals of the Association of American Geographers*) as well as other development-oriented journals (e.g. *Development and Change, The Ecologist, World Development, Society and Natural Resources, Capitalism, Nature, Socialism*). Thus, Third World political ecology is a research field that lacks a major journal of its own with the result that political ecologists lack a reliable reference point for current developments and debates within the field. In contrast, virtually all of the environmental research fields described in Table 1.2 possess at least one major journal of their own. Hence, for example, the existence of such journals as *Cultural Ecology* (1971), *Environmental Politics* (1992), *Environmental Management* (1976), *Journal of Environmental Management* (1973), *Human Ecology* (1973), *Environmental History Review* (1976), *Environmental History* (1995), *Ecological Economics* (1989) and *Global Environmental Change* (1991).

Third, Table 1.2 suggests that Third World political ecology has stronger potential or actual affinities with certain fields than with other fields. The long-standing link with cultural ecology (ecological anthropology) is a case in point. Although Third World political ecology developed in part as a critique of cultural ecology (see Table 1.1), the latter has moved on from the cybernetics of the 1960s and 1970s, and has developed a much more sophisticated appreciation of culture–environment interactions (e.g. Ingold, 1992). That appreciation – significantly – addresses questions of power in considering the relationships between environment and culture (Croll and Parkin, 1992). As such, recent research in cultural ecology converges with that in Third World political ecology around the role of power relations in conditioning human–environmental interaction.

To some extent, there is also a convergence taking place between Third World political ecology and environmental history. Thus, while the latter has begun to accord greater attention than in the past to understanding Third World environmental change (e.g. Beinart and Coates, 1995; Rangarajan, 1996), the former has drawn on insights from environmental history to acquire a 'much-needed historical depth' (Peet and Watts, 1993: 241) previously often lacking in the field. However, the affinity between these two environmental research fields must not be exaggerated as many environmental historians appear not to share the radical perspective dear to most political ecologists.

In contrast, Third World political ecology would appear to have a potential affinity with the field of ecological economics precisely because the two share a radical perspective on the global capitalist system, and the futility of all attempts to render that system environmentally sustainable through reformist measures. However, ecological economists have scarcely begun the immense task of effecting a 'radical' integration of economy and ecology, although promising work in this direction has already been undertaken (e.g. Martinez-Alier, 1990; O'Connor, 1994a).

Finally, Table 1.2 suggests that Third World political ecology is ecocentric (not technocentric) in its outlook on human–environmental interaction. This book has already suggested that the radical content of Third World political ecology distinguishes it from most other fields that address Third World environmental problems. Yet the field of political ecology nonetheless shares with cultural ecology and human ecology an ecocentric outlook that is founded on a distrust of technologically based reformist solutions to Third World environmental problems, and there tends also to be agreement on the need for local-level management responses to those problems. Although rarely stated, the lodestar of most political ecologists is a relatively decentralised political system that blends socialism and anarchism in the pursuit of social justice (see Introduction). Political ecologists are thus in the main ecocentrics who are politically to the left, and who seek to reconcile material equity considerations and environmental conservation in the context of calls for an end to the global capitalist system (cf. Pepper, 1993; Norgaard, 1994). Table 1.3 sets out the main differences between ecocentrism and technocentrism to illustrate the ecocentric bases of Third World political ecology.

This table (in conjunction with Table 1.2) shows the wide gap that exists between Third World political ecology and selected other research fields. Environmental economics, global ecology and environmental management, for example, are based largely on a technocentric outlook which makes assumptions about precisely those matters (e.g. economic growth, political-economic structures) that political ecologists argue ought

Table 1.3 Environmental outlooks: technocentrism and ecocentrism

Technocentrism	*Ecocentrism*
1 Modified sustainable economic growth	1 Limits to, and undesirability of, economic growth
2 Large role for technological development as a provider of solutions for environmental problems	2 A distrust of science and technological fixes
3 Environmental solutions can co-exist with existing social and political structures	3 Radical social and political change necessary. A preference for decentralised social and political organisation
4 Anthropocentrism and a commitment to intra-generational and inter-generational equity	4 Intrinsic value of nature or at least a weaker version of anthropocentrism; a commitment to social justice within human society and between humans and non-human nature

Source: Adapted from Garner, 1996: 30

to be at the heart of environmental research. Thus, although at first glance it might appear that there are complementarities between Third World political ecology and these three fields, divergent environmental outlooks tend to place them in opposing camps.

The preceding discussion has not attempted a comprehensive treatment of the different environmental research fields, but rather has sought to show that Third World political ecology relates to other environmental research fields in complex ways. Specifying precise research affiliations and locations is a hazardous endeavour in a context of rapid intellectual change in all of these fields. What is clear from this brief discussion, however, is that Third World political ecology holds a distinctive place in social-science research on environmental matters as a result of its concern to integrate the 'concerns of ecology and a broadly defined political economy' (Blaikie and Brookfield, 1987: 17) in a Third World context. This research is undoubtedly influenced by research in contiguous fields, but is also shaped by the specific approaches that political ecologists adopt within the field itself.

APPROACHES

Political ecologists share a broadly similar political-economy perspective, but adopt a variety of approaches in applying that perspective to the investigation of human–environmental interaction in the Third World. This variety of approaches reflects, in turn, differing research priorities within the field.

Figure 1.1 summarises the main approaches adopted by political ecologists to illustrate the different ways in which research has been conducted in the field. The point here is not to suggest that all scholarship in Third World political ecology fits neatly within one or the other category. Indeed, in many cases scholars have combined elements from two or more approaches (e.g. problem-region). Nor is it to argue that one approach is necessarily better than another approach (although we provide a justification for this book's choice of an actor-oriented approach) since the choice of an approach will depend on the questions that the researcher wishes to address. Rather, Figure 1.1 is designed to explore the ways in which political ecologists have approached the subject through a 'plurality of purpose and flexibility of explanation' (Blaikie and Brookfield, 1987: 25).

One approach has been to orient research and explanation in Third World political ecology around a specific environmental problem or set of problems such as soil erosion, tropical deforestation, water pollution or land degradation. This approach constitutes in many respects a 'traditional' geographical research theme associated with understanding the human impact on the physical environment (Goudie, 1993), but with a distinctive political-economy twist. Geographers have been predictably at the forefront of research adopting this approach. Thus, Blaikie's (1985) study on the political economy of soil erosion explained this problem in terms of a hierarchy of inter-linked social, political and economic forces operating at local, regional and global scales. Blaikie and Brookfield (1987) elaborated this work with reference to the more all-encompassing problem of land degradation. Other scholars have focused on specific environmental problems in various parts of the Third World: tropical deforestation in Brazil (Hecht and Cockburn, 1989) or Indonesia (Dauvergne, 1993/4), water scarcity in Botswana (Peters, 1984), or rangeland degradation in Sahelian Africa (Turner, 1993).

A second approach involves focusing on a concept that is perceived as having important links to political-ecology questions. To understand the latter is partly to appreciate the ways in which ideas are developed and understood by different actors, and how attendant discourses are developed to facilitate or block the promotion of a specific actor's interests (Escobar, 1996). As Schmink and Wood (1987: 51) observe, 'ideas are never "innocent" ... [they] either reinforce or challenge existing social and economic arrangements'. Political ecologists have thus explored the implications of the 'dominant discourse of scientific forest policy' in various settings (e.g. Peluso, 1992; Jewitt, 1995; Bryant, 1996a). Comparable work has been done on 'soil erosion and conservation' discourses (Beinart, 1984; Blaikie, 1985; Zimmerer, 1993). Other scholars have evaluated the social construction of natural hazards, disasters and vulnerability (e.g. Watts and Bohle, 1993; Blaikie et al., 1994), building on the earlier work of Wisner and O'Keefe (see above). However, it is work on sustainable or green development which perhaps best represents research adopting this approach. Thus, Redclift's (1987) analysis of sustainable development examined the possible contradictions of the quest to reconcile environmental conservation

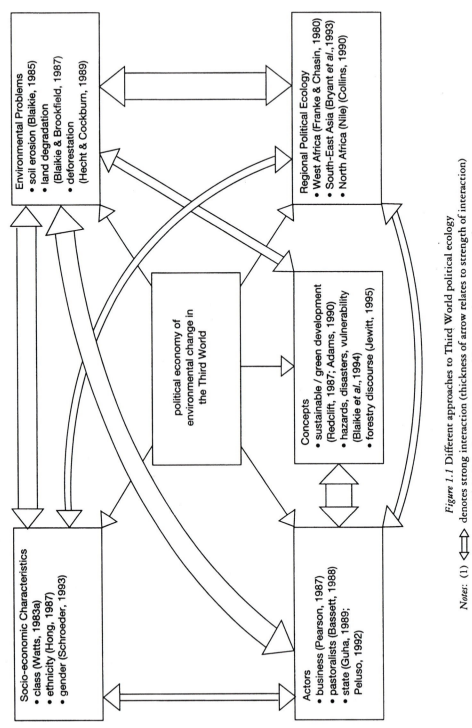

Figure 1.1 Different approaches to Third World political ecology

Notes: (1) ⟺ denotes strong interaction (thickness of arrow relates to strength of interaction)
(2) Indicated sources may combine more than one approach

and economic development within the contemporary global capitalist system. Adams (1990) investigated further whether green development called for a reformist or radical political strategy, and the policy implications of differing strategic choices. Escobar (1996: 48), meanwhile, explores the influential discourse surrounding sustainable development in order to clarify 'dominant assumptions about society and nature, and the political economy that makes such assumptions possible'.

A third approach is to examine inter-linked political and ecological problems within the context of a specific geographical region. This 'regional political ecology' has reflected a concern to take into account 'environmental variability and the spatial variations in resilience and sensitivity of the land [sic]' as well as 'theories of regional growth and decline' (Blaikie and Brookfield, 1987: 17). Franke and Chasin (1980) and Watts (1984) in West Africa, McDowell (1989) and Bryant et al. (1993) in South-East Asia, and Waterbury (1979) and Collins (1990) in North Africa (i.e. the Nile river) illustrate approaches that frame political-ecological questions in a regional context. The nature of the problems often varies from region to region, but a shared goal is to evaluate those problems in a regional context.

A fourth approach is to explore political-ecological questions in the light of socio-economic characteristics such as class, ethnicity or gender. As noted above, Third World political ecology's structural legacy has been associated with work that sought to explain local conflict in terms of capitalist relations of production and class conflict. Cliffe and Moorsom (1979) and Watts (1983a) on Africa, and Bunker (1985) and Branford and Glock (1985) on Latin America, highlight political-ecology research based on class analysis. Other scholars have undertaken research linked to non-class factors. For example, Hong (1987) and Colchester (1993) examine the impact of state-sponsored logging on indigenous peoples in Sarawak (Malaysia), as do Smith (1994) and Bryant (1996b) with regard to ethnic minority groups in Burma. This work shows how politically and economically marginal ethnic minorities (often linked to shifting cultivation or hunting-gathering practices) often bear the brunt of the costs associated with environmental degradation. In contrast, other work shows how household relations are a 'deeply contested terrain' (Watts, 1989: 12) in which men and women struggle for control of environmental resources (notably land), capital and labour. Dankelman and Davidson (1988), Leach (1991), Schroeder (1993), Joekes et al. (1995), and Rocheleau and Ross (1995) reflect a gendered approach with reference to different geographical regions or environmental problems. What has been termed 'feminist political ecology' (Rocheleau et al., 1996) thus demonstrates how gender inequalities relate to environmental change and conflict.

Finally, a fifth approach emphasises the need to focus on the interests, characteristics and actions of different types of actors in understanding political-ecological conflicts. An actor-oriented approach seeks to understand

23

such conflicts (cooperation too) as an outcome of the interaction of different actors pursuing often quite distinctive aims and interests (Long and Long, 1992). Thus, Bassett (1988) and Toure (1988) explore the conflict between pastoralists, farmers and the state in Ivory Coast and Senegal that is associated with the divergent interests of these actors. Pearson (1987), Leonard (1988), Morehouse (1994) and Moody (1996) consider the links between transnational corporate strategies, environmental degradation and local political conflict in such places as Mexico, Papua New Guinea, Namibia and India. Guha (1989), Peluso (1992) and Bryant (1997a) examine how forest politics in India, Java (Indonesia) and Burma has been conditioned by the struggle of different actors (peasants, shifting cultivators, businesses) with the state, as well as struggles within the state between departments (agriculture, forestry), pursuing different yet overlapping mandates. The actor-oriented approach, then, relates an understanding of actors to political and ecological processes.

It needs to be reiterated that these different approaches to Third World political ecology are not mutually exclusive. Indeed, as noted, scholars frequently combine approaches or use different approaches in different pieces of research. All of the approaches outlined above represent potentially useful ways of approaching the subject-matter depending on the research priorities of the individual scholar. Nonetheless, the decision to adopt an actor-oriented approach in this introduction to Third World political ecology merits comment here.

Our choice has been guided in the first instance by Moore's (1993, 1996) repeated claim that there has been insufficient 'politics' in Third World political ecology. The Introduction suggested that, although this accurately described much research in the late 1970s to mid-1980s, the claim was fast becoming outdated in the mid-1990s. Recent research (Neumann, 1992; Peluso, 1993a, 1995; Rocheleau and Ross, 1995) has shed new light on the 'micro-politics' that informs environmental conflict and cooperation at the local level. Yet there is also a danger, if such research becomes the norm, that we will lose sight of 'the forest for the trees'. Thus, micro-political research tells us much about the local dimensions to political ecology, but is less informative about the wider theoretical and comparative significance of the types of actors involved in local interaction. Yet if Third World political ecology is about the struggle between actors for control over the environment, it is as vital to appreciate the wider impact and significance of actors as it is to understand how those actors may interact in a given locality.

The primary aim of this book is thus to introduce the reader to the field of Third World political ecology in terms of a general evaluation of the political role of different types of actors in human–environmental interaction in the Third World. There are several advantages to this approach. First, by examining the general role and significance of selected actors in Third World environmental change, we are able to situate the findings of much local-level empirical research in theoretical and comparative perspective. In doing so, we avoid the 'Achilles heel of localism' (to adapt an expression used by Esteva and Prakash, 1992) that bedevils much work in the field – that is, the sense that research findings have limited applicability outside of the locality in which they were derived.

Second, by integrating theoretical and comparative insights as to the role and significance of different actors, we seek to provide a reasonably comprehensive picture of the motivations, interests and actions of those actors in a manner that is not possible through locality studies. The latter typically describe only those traits of actors relevant to understanding developments at the chosen locality. In contrast, our objective is to attempt a well-rounded assessment of different types of actors, including notably their political strengths and weaknesses in relation to other actors. In the process, we seek to go beyond the stereotypes that bedevil much environmental research, including Third World political ecology. Thus, descriptions of ecologically 'predatory' states and transnational corporations, 'eco-friendly' non-governmental organisations or grassroots actors (e.g. poor farmers or shifting cultivators) are common in the literature, but tend to obscure the complexities and contradictions associated with the actions of all actors. Our goal is to provide a more reasoned appreciation of those complexities and contradictions than has hitherto been the case in Third World political ecology.

Finally, by emphasising the role and interaction of actors in environmental conflict in the Third World, we reiterate the central importance of politics in political ecology. It seems to us that there are two things that are at the heart of any meaningful understanding of politics: (1) an appreciation that politics is about the interaction of actors over environmental (or other) resources; (2) a recognition that even weak actors possess some power to act in the pursuit of their interests. The former point suggests that politics is a process in which actors partake and play a central role; indeed, that the interaction of actors is the very stuff of politics (cf. Hurrell and Kingsbury, 1992). This point has not always been clearly acknowledged by political ecologists, especially those whose work in the past has been prone to economic reductionism. The second point elaborates this suggestion by signalling our belief that Third World political ecology must continue its movement away from 1980s structuralism and towards a full appreciation of the role of agency in human affairs (Redclift, 1992). Giddens's (1979: 149) reminder that 'all power relations, or relations of autonomy and dependence, are reciprocal: however wide the asymmetrical distribution of resources involved, all power relations manifest autonomy and dependence "in both directions"' is apposite here, and will be explored in greater detail in Chapter 2. The point is not, of course, to deny the impact on actors of global processes – indeed, we argue in this book that Third World political ecology must vigorously engage in debates over 'globalisation' (Redclift and Benton, 1994) and the role of the Third World generally in global political and economic processes (Miller, 1995). Rather, it is to assert the need to ground an understanding of global (and regional or local) processes in an appreciation of the role of specific actors in their development – and thereby to render these processes simultaneously more tangible and more meaningful in political terms (Blaikie, 1995a).

The remainder of this book explores the key types of actors involved in environmental change and conflict in the Third World to that end. Thus, successive chapters examine the role of states, multilateral institutions, businesses, environmental non-governmental organisations and grassroots actors in such change and conflict. We make no attempt to cover all possible types of actors in what follows, but do seek to account for what we consider to be the most important types at the present juncture. However, before turning to the discussion of actors, it is necessary first to examine the nature of the contemporary environmental crisis in the Third World, and the role of political ecology in its possible interpretation through the idea of a politicised environment.

2

A POLITICISED ENVIRONMENT

The prevailing image of the Third World today in the development literature is that of a part of the world trapped in perennial crisis. Environmental considerations figure prominently in this literature as concern grows over such problems as tropical deforestation, soil erosion or desertification. Curiously lacking from much of this writing, however, is any sense of the political and economic factors that contribute to the Third World's environmental problems. To be sure, the link between poverty and environmental degradation is now widely recognised as 'poverty's profile ... has become increasingly environmental' (Durning, 1990: 135). Yet the story typically ends there with the result that the political and economic forces that contribute to the 'desperate ecocide' of the poor (Blaikie, 1985) remain largely unexplored (e.g. World Commission on Environment and Development, 1987; World Bank, 1992).

In contrast, political ecologists attempt to complete the story through work oriented around the idea of a 'politicised environment'. In the process, they agree with their mainstream counterparts that the Third World is wracked by an environmental crisis but, unlike the latter, then ask 'Whose environmental crisis?' That question focuses attention squarely on issues of political and economic causality, and ensures that political ecologists do not fall into the trap (as do many mainstream writers) of treating the Third World's 'environmental crisis' in isolation from the much wider developmental crisis to which it is inextricably linked. This book approaches the Third World's environmental crisis from the vantage-point of the interests and actions of the main actors involved in the conflict that habitually surrounds that crisis. However, this chapter first examines the idea of a politicised environment – its possible meanings and topography – as well as the implications for the field of Third World political ecology of research based on this idea.

DIMENSIONS

Central to the idea of a politicised environment is the recognition that environmental problems cannot be understood in isolation from the political and economic contexts within which they are created. Thus, to describe problems of, say, desertification, tropical deforestation, soil erosion or wildlife depletion is simultaneously to consider the political and economic processes that generate or exacerbate those problems. Indeed, the very definition of an environmental 'problem', and the priority attached to that problem by society, may itself be a reflection of those same processes (Hoben, 1995; Rocheleau *et al.*, 1995; and see below). Harvey's (1993: 25) observation, noted earlier in this book, that 'all ecological projects (and arguments) are simultaneously political-economic projects (and arguments) and vice versa', is apposite here.

A sense of what it means to conceive of a politicised environment in the Third World can be gained from considering briefly how mainstream scholars attempt to understand environmental change without reference to political and economic processes. On the one hand, these scholars tend to describe such change in relation to sweeping historical narratives that usually feature accounts of population growth and intensifying per capita human impacts on the environment (e.g. Ehrlich and Ehrlich, 1990; Goudie, 1993). These narratives tend to operate at a high level of abstraction with the inevitable result that, if the issue of 'blame' for the Third World's environmental crisis is ever raised, it is typically apportioned uncritically to 'humankind' as a whole. On the other hand, mainstream scholars who consider the crisis in a less abstract light tend to emphasise 'technical' dimensions as part of a managerial 'problem-solving' approach. Yet as Blaikie (1985: 5) notes with reference to soil conservation programmes, an emphasis on technical solutions leads typically to policy failure, which, in turn, is associated with the general unwillingness to 'make explicit more fundamental assumptions' about the political economy of environmental change.

In contrast, political ecologists start from the premise that environmental change is not a neutral process amenable to technical management. Rather, it has political sources, conditions and ramifications that impinge on existing socio-economic inequalities and political processes (Bryant, 1992). Several linked assumptions inform work by political ecologists. First, they accept the idea that costs and benefits associated with environmental change are for the most part distributed among actors unequally. To be sure, selected physical changes linked notably to the nuclear, chemical and biotechnology industries may involve the generation of costs (risks) to which all are equally exposed (but note that this does not necessarily apply to the benefits associated with these changes). The implications of what Beck (1992) terms the 'risk society' have yet to be adequately considered by political ecologists

28

dealing with Third World environmental problems, a point returned to in the Conclusion of this book. Yet, however momentous the nature of these generalised costs (risks) may be, it is nonetheless the case that the costs associated with the vast majority of types of environmental change that affect the Third World today are distributed far from equitably.

Second, political ecologists assume that an unequal distribution of environmental costs and benefits reinforces or reduces existing social and economic inequalities. This assumption reiterates the point that environmental and developmental concerns are inseparable – and that any change in environmental conditions must affect the political and economic status quo, and vice versa. If the environment in the Third World is above all a 'livelihood issue' (Redclift, 1987) then any change to that environment will inevitably alter the ability of different actors to earn a livelihood.

Finally, political ecologists argue that the differentiated social and economic impact of environmental change also has political implications in terms of the altered power of actors in relation to other actors. Thus, environmental change not only signifies wealth creation for some and impoverishment for others, it also thereby alters the ability of actors to control or resist other actors. A striking illustration of this point relates to the ubiquity of conflict over environmental resources in the Third World. The existence of such conflict highlights the importance that diverse actors attach to those resources, as well as their recognition that changing environmental conditions hold political (as well as economic) opportunities and consequences.

Political ecologists use these assumptions to interpret the Third World's politicised environment. But how is that politicised environment itself to be understood? The conventional way is to think in terms of discrete types of environmental change – deforestation, soil erosion, flooding, salinisation, etc. – and to relate them separately to political and economic processes. Yet such an approach tends to perpetuate precisely the nature–society dichotomy that some radical scholars accuse their mainstream counterparts of uncritically helping to sustain (Szerszynski et al., 1996). We would suggest that a way around this impasse is to focus not on a description of the physical environmental changes themselves, but rather on the way in which those changes relate to human activities.

In this spirit, we would suggest that there are three dimensions to a politicised environment: (1) everyday; (2) episodic; and (3) systemic. These dimensions are set out in Table 2.1 in relation to physical changes, the rate of impact, the nature of the human impact, the political response and key concepts. The first 'everyday' dimension involves physical changes (e.g. deforestation, salinisation) that simultaneously derive from day-to-day human practices and unequally affect those same practices on a daily basis. The second 'episodic' dimension comprises physical changes (e.g. flooding, high winds/storms) that often have a massive, immediate and highly unequal human impact, but occur sporadically over time and are frequently described, usually inaccurately, as 'natural' disasters. The third 'systemic' dimension involves physical changes that derive from industrial activities (e.g. nuclear fallout, pesticide concentrations in the human food chain) which are potentially 'equal' in their human impact. Beck (1992) has elaborated this latter dimension through the notion of a 'risk society', but the literature surrounding this notion has been largely concerned with its applicability (or otherwise) in a First World context. It is far from clear, moreover, how equal in practice these systemic changes are when the Third World context is considered (e.g. Bull, 1982 on pesticide poisoning among the Third World poor). However, most political ecologists have yet to explore systematically these latter issues, and have instead focused their attention on the first and second dimensions of a politicised environment. This book is also consequently mainly concerned with understanding actors in relation to the first two dimensions, although we touch on questions relating to the third dimension from time to time (e.g. Chapter 5) and return to it again in the Conclusion.

Table 2.1 Dimensions of a politicised environment

Dimensions	Physical changes	Rate of impact	Nature of human impact	Political response	Key concept
Everyday	Soil erosion, deforestation, salinisation	Gradual and may not even be perceived for a long time	Cumulative and typically highly unequal; the poor are the main losers	Livelihood protests / resistance	Marginality
Episodic	Flooding, high winds / storms, drought	Often sudden but occasionally drawn out	May have general impact but unequal exposure means that the poor are the main losers	'Disaster' relief	Vulnerability
Systemic	Nuclear fallout, pesticide concentration, biologically modified species	Gradual and not necessarily perceived but also potentially sudden	Tends to have a general impact	Popular distrust of official 'experts'	Risk

Here it is thus important to consider further the everyday and episodic dimensions of a politicised environment set out in Table 2.1. The first thing to note is that there is nothing especially 'natural' about either dimension. This claim may at first glance appear unfounded since the environment has long changed independent of human design or influence – viz. soil depletion, forest change, floods or earthquakes. Yet a variety of factors suggest the need for a more complex and anthropocentric appreciation of environmental change. Thus, as Blaikie (1985) and Blaikie and Brookfield (1987) notably suggest, physical environmental change only becomes meaningful to humankind when it is perceived to be a problem or an opportunity by different actors. Indeed, in certain cases (e.g. soil erosion) environmental change may long go unnoticed by those actors – and thus, will not figure in any significant way in human affairs at all.

Beyond the question of human perceptions, however, is the issue of human-induced environmental change which is assuming an ever greater role in the ultimate rate and extent of overall environmental change. It would, of course, be wrong to equate environmental change with human-induced environmental change or, for that matter, to underestimate the unpredictability of ecological processes themselves (Zimmerer, 1994). Yet a defining trait of human development has been the growing ability to manipulate, if not control, physical or biological processes – 'produced nature' as it were (Castree, 1995). In the measure that produced nature replaces 'original' nature, then, speaking about natural processes or disasters becomes increasingly misleading.

However, as O'Keefe et al. (1977) pointed out long ago, disasters are unnatural for another reason relating to the spatial impact on human settlements of episodic changes. Thus, the distribution of costs involved in most episodic changes is not random. Rather, it is the poor and otherwise marginalised members of society who are disproportionately affected by flooding, drought or earthquakes. The 'disaster-proneness' of these actors is much greater than is the case with wealthier or traditionally more powerful actors (Blaikie et al., 1994). For these diverse reasons, therefore, it is possible to speak of the 'unnaturalness' of most forms of everyday and episodic environmental change.

A second thing to note about the everyday and episodic dimensions of a politicised environment is their interconnectedness. It is true that debate among scholars continues as to whether everyday and episodic physical environmental changes can be linked together. A case in point is the widely reported and believed link between upland deforestation and lowland flooding, the assumption being that the deforestation of critical watersheds is associated inevitably with intensified flooding events. However, evidence from the Himalayas casts doubt over such an equation, suggesting rather that flooding events are related primarily to rainfall and other factors (Ives and Messerli, 1989).

Yet this debate on the physical connection of the everyday and episodic dimensions should not obscure the existence of social connections – that is, the human activities that link together these two dimensions. Thus, a key social factor associated with everyday physical changes is that of marginalisation (Blaikie and Brookfield, 1987). This process occurs when poor grassroots actors such as farmers or shifting cultivators are pushed onto lands that are economically marginal as a result of their marginal political and economic status. Desperate to extract a living from such lands, these actors intensify production, but in the process often only increase the land's ecological marginality (i.e. reduced capability). This vicious cycle continues since the prospect of an actor deriving a livelihood from the land is thereby diminished. The result is that 'land degradation is both a result of *and* a cause of social marginalization' (Blaikie and Brookfield, 1987: 23).

In contrast, an important social factor in understanding the relationship between episodic physical change and human activities is that of vulnerability. The latter constitutes a measure of the exposure of different actors in society to episodic changes such as flooding or high winds/storms. However, the relative vulnerability of actors to such changes is often related to their degree of political and ecological marginality. Thus, marginalised farmers working ecologically marginal lands are typically more vulnerable to drought or pestilence than wealthier or more powerful farmers who work better lands. These episodic changes may, in turn, reinforce even further the marginality of these poor actors by leaving them physically weaker, and thus less able to work the land, or by forcing them into an increased dependency on local money-lenders or politicians. It is thus, above all, in relation to human activities that the connections between the everyday and episodic dimensions of a politicised environment may be most clearly seen.

A third, and related point about these two dimensions is that both everyday and episodic dimensions are closely linked to the intensification of social and economic inequalities. On the one hand, everyday and episodic physical changes are both in different ways related to existing socio-economic inequalities. Everyday changes are linked, for example, to unequal land holdings which might be associated with the need for actors marginalised from good lands to degrade the ecologically marginal lands under their control in order to survive (Broad, 1994). Similarly, as noted above, the costs associated with episodic changes are typically borne disproportionately by these same marginalised actors. On the other hand, everyday and episodic physical changes also tend to reinforce existing social and economic inequalities in so far as they diminish even further the resources and power of weaker actors while they simultaneously strengthen the power and resources of stronger actors. Thus, intensified social and economic disparities are often associated with the everyday and episodic dimensions of a politicised environment.

A fourth thing to note about these two dimensions is their relationship to the political process. Everyday and episodic physical changes may reflect and strengthen the political control of traditionally powerful actors (landlords, state officials, large corporations) over their weaker counterparts in a manner similar to that associated with socio-economic inequalities. However, such changes may also prompt a response by weaker actors through various forms of covert or overt resistance. The ensuing conflict may lead to a partial reversal of the socio-economic inequalities discussed above, thereby highlighting that everyday and episodic changes do not necessarily result in a perpetuation of the political and economic status quo (see below and Chapter 7).

The preceding discussion has outlined selectively the different dimensions of a politicised environment. It has suggested that different types of environmental change become meaningful only in the context of an integrated understanding of human–environmental interaction in which political and economic inequalities influence the social distribution of the costs and benefits of everyday and episodic changes. The discussion so far has shown that the Third World's environmental crisis is more of a crisis for some actors than it is for other actors. Indeed, it has even suggested that this crisis might be a source of opportunity for powerful actors. This point becomes clearer when we consider the question of scale in a politicised environment.

SCALE

To think about a politicised environment is to reflect on how that environment is constituted and how it changes at different scales in relation to both physical problems and actors. There is a growing literature on the nature, extent and possible human significance of diverse environmental problems as they manifest themselves at the local, regional or global scale (Goudie, 1993; Pickering and Owen, 1994). Different physical problems are frequently associated primarily with one or another scale (although they also can operate at different scales). Thus, for example, soil erosion or deforestation are essentially seen as being local problems, while coastal pollution or drought may be described as regional problems. In contrast, greenhouse warming and ozone depletion are considered to be classic global problems.

However, the existence of a whole host of environmental problems at different scales cannot be adequately understood without recognising simultaneously that different actors contribute to, are affected by, or seek to resolve, environmental problems at different scales. Further complexity derives from a situation in which actors become differentially involved at any given scale. Thus, while one actor's involvement may reside predominantly in contributing to an environmental problem, another actor might largely be involved in its attempted resolution, and a different actor again may be primarily involved only because it is particularly affected by the problem. In reality, actors typically become involved in an environmental problem on several or even all counts.

To illustrate this argument, Table 2.2 sets out possible links between scale, environmental problems and actor involvement with reference to deforestation (local), drought (regional) and greenhouse warming (global). This table is designed only to highlight general issues, some of which are recurrent themes in Third World political ecology, while others have been scarcely addressed by the field. However, most issues will feature at various stages in this book as the focus shifts to the role of specific actors in environmental problems and conflict.

A first issue raised by Table 2.2 relates to the distribution of the costs and benefits associated with environmental problems at different scales. Not surprisingly, it is the relatively poor and weaker grassroots actors (e.g. farmers, shifting cultivators) who bear a disproportionate share of the direct costs (impact) associated with local (deforestation) and regional (drought) environmental problems compared with their wealthier or more powerful counterparts (e.g. states, businesses, multilateral institutions). However, the former are often doubly disadvantaged in that they rarely receive a significant proportion of the benefits that usually attach to the economic activities that contribute to local or regional problems in the first place. In contrast, powerful actors derive their position in society in part from activities such as large-scale logging or cash-crop production that may be the main human contribution to such environmental problems. These same actors are often able to escape most of the costs associated with the human impact of these problems. To be sure, it needs to be reiterated that weaker actors may receive some benefits from activities that contribute to deforestation or drought, while more powerful actors rarely escape completely from paying some of the costs associated with these problems. Yet much work in Third World political ecology supports this argument that costs fall mainly on poorer and weaker actors while benefits accrue mainly to wealthy and more powerful actors (e.g. Horowitz and Little, 1987; Hecht and Cockburn, 1989; Thrupp, 1990; Stonich, 1993).

The distribution of costs and benefits becomes more complicated at the global scale. Yet global environmental problems do not relate to human activities in an undifferentiated manner. On the one hand, the human impact of greenhouse warming (e.g. coastal flooding, altered agricultural productivity) is likely to be complex, but will nonetheless tend to hit weaker actors the hardest due to their inability to respond flexibly to new social and ecological situations. On the other hand, the benefits associated with economic activities that are the main contributors to this environmental problem have largely accrued to powerful actors located primarily in the First World. What Agarwal and Narain (1991) term 'environmental colonialism' is reflected in highly unequal use of global energy and natural resource supplies (and inevitably the world's atmosphere as pollutants are

Table 2.2 Scale, environmental problems and actor involvement

Actor	Local scale (deforestation)			Regional scale (drought)			Global scale (greenhouse warming)		
	Contribution (and potential benefits)	Impact (costs)	Resolution	Contribution (and potential benefits)	Impact (costs)	Resolution	Contribution (and potential benefits)	Impact (costs)	Resolution
States	Logging by state agencies; policies that encourage deforestation by others (e.g. business logging)	Potential revenue loss	Typically the leading actor (i.e. reforestation policy)	Policies promoting large cash-crop production operations linked to human-induced soil degradation	Revenue loss, social unrest; politically severe impact for small states hard hit by the problem	Key role (with other actors) in devising new land-management practices	State manufacturing and energy production; industrial policies encouraging business	Potentially great impact for poor low-lying states	Cooperation and agreements between states
Grassroots actors (i.e. poor farmers, shifting cultivators, nomadic pastoralists)	Permanent and/or temporary forest clearance	Loss of needed forest products; loss of livelihood; elimination of way of life	Usually excluded from official efforts, but 'participation' occasionally sought (i.e. social forestry)	Over-exploitation of marginal lands (although this may be related to marginal status of these actors)	Often devastating (including death or destroyed livelihoods)	Usually excluded but occasional 'participation' sought today	Relatively small	Potentially great for those worst affected by altered ecological conditions but unable to respond	Excluded
Business	Small- to large-logging operations	Often negligible as move to new logging concessions	Growing role via plantation forestry	Large cash-crop production operations linked to soil degradation	Potentially reduced profits	Investment to restore operations but on 'sustainable' basis; contract farming	Manufacturing and energy production; avoidance of pollution controls	Not yet but potentially reduced profits as regulation increases	Voluntary guidelines; respond to state policies
Multilateral institutions (technical- or finance-orientated)	Technical advice and consultancy; loans to sustain economic activities; structural adjustment lending	Limited to criticism of their role	Technical advice and consultancy; loans supporting 'green' activities	Technical advice and consultancy; loans supporting export-orientated cash cropping; structural adjustment lending	Criticism only	Technical advice and support, and loans to support new management practices	Advice and loans to states and industry	Limited criticism	Input in global negotiations between states
Environmental non-governmental organisations	None	No costs but may increase funding available to this actor	Support grassroots actors; lobby other actors (especially the state)	None	No costs but increased revenue due to 'crisis' situation	Potentially leading role	None	No costs (may slightly increase funding)	Public education campaigns; lobby states and businesses

released into the air) by the First and Third Worlds. Indeed, even in the Third World itself, a substantial (albeit unspecified) proportion of activities that contribute to greenhouse warming 'arise essentially not out of Third World consumption but Western consumption' (Agarwal and Narain, 1991: 24). The economic success of selected Asian countries is beginning to change this highly unequal situation, yet there is still a long way to go before the Third World matches the First World's historic and contemporary contribution to global warming. Further, within the successful Asian countries themselves, the economic benefits from polluting activities accrue disproportionately to the newly prosperous middle classes.

A second issue arising from Table 2.2 concerns the role of different actors in efforts to resolve environmental problems at the local, regional or global scale. The state tends to play a pivotal role in efforts to resolve problems at all three scales, reflecting this actor's 'stewardship' role in society. At the global scale, states have found cooperation with each other imperative to attempt a global approach in dealing with environmental problems that transcend the ability of any one state to address (see Chapter 3). Yet states also receive the technical and financial support of multilateral institutions such as the World Bank or Food and Agriculture Organisation to introduce new policies at different scales (see Chapter 4). A growing trend in the late twentieth century has also been for states to seek the support of business (especially transnational corporations) in attempting to combat environmental problems at various scales. The prospect of a 'partnership' between these two types of actors was enhanced as a result of the Rio Earth Summit in June 1992, which specifically acknowledged the contribution of business in solving the world's environmental problems (see Chapter 5).

The interests and concerns of these three different types of actor may vary, but they nonetheless share two things in common that are of interest to the present discussion. They all adopt a technical problem-solving approach to environmental problems which relies heavily on the input of First World science and professional expertise. The tendency has been for these groups to abstract environmental problems (e.g. deforestation, drought, greenhouse warming) from the political and economic contexts within which they have been created or exacerbated, and then to propose generic solutions based on standardised assessments of the problems. The 'deafening silence' of states, multilateral institutions and businesses when it comes to questions of political and economic causation is related, in part or in whole, to a second factor that these actors share in common – namely, a close association with economic activities that have been major contributors to environmental problems at the local, regional and global scales. The direct and/or indirect contribution of states, businesses and multilateral institutions to such activities as large-scale logging, cattle ranching or cash-crop production is considered in subsequent chapters of this book. However, the point here is simply to note that the actors who have often played a major role in contributing to the Third World's environmental problems are also largely in charge (formally at least) of devising solutions to those problems.

In contrast, those actors that have been typically worst affected by environmental problems, especially at the local or regional scale, have so far played a marginal role in these problem-solving efforts. That weaker grassroots actors such as poor fishers or farmers have been relegated generally to such a role is hardly surprising given the discussion of the previous section. Those actors whose environmental management role has been characterised by political, economic and ecological marginality up until now can expect, all things being equal, to play a similarly marginal role in the attempt to 'solve' the Third World's environmental problems. Thus, just as there is a mismatch between which actors bear the costs of, and which actors benefit from, the economic activities associated with environmental problems at various scales, so too there is an asymmetry between which actors bear the costs of, and which actors largely control, efforts to solve the problems.

The latter asymmetry is no idle matter as it has important implications for the future of environmental management, and which interests are served by such management. Thus, for example, states, businesses and multilateral institutions have responded to the problem of tropical deforestation through the promotion of a system of large-scale commercial plantation forestry. However, while the creation of vast eucalyptus or pine plantations (used in the pulp and paper industries) may 'green' selected areas in a commercially 'useful' manner, these plantations have become the centre of intense conflict as local farmers (often supported by environmental non-governmental organisations or ENGOs) resist a 'solution' that often seems only to result in local land degradation and depressed local livelihoods (Marchak, 1995; Lohmann, 1996; and below).

A third issue linked to Table 2.2 concerns the role of grassroots actors and ENGOs in the evolution of environmental problems at various scales. The discussion so far has suggested that the preeminent role played by contextual actors (states, multilateral institutions, large businesses) in that evolution, and their ability to influence patterns of human–environmental interaction more or less as they wish, is to their ultimate benefit. Yet, as we suggested in Chapter 1 and reiterate below, power relations may be highly unequal, but they are rarely, if ever, 'one-way'. Thus, and as Chapter 7 explores in detail, relatively weak grassroots actors (e.g. poor farmers, shifting cultivators, nomadic pastoralists, fishers) nonetheless have 'weapons of the weak' (Scott, 1985) that ensure that, while these actors may be neglected, they can rarely be ignored altogether by their more powerful counterparts. It is partly for this reason that states and other traditionally powerful actors have felt a growing compulsion to consult with (if not actually listen to) grassroots actors, especially over environmental problems at the local scale (Peluso, 1995; Braganza, 1996).

A further factor fuelling this 'consultation process' relates to the rise of ENGOs in recent years as prominent actors in their own right in the Third World's environmental conflicts (see Chapter 6). As Table 2.2 shows, this type of actor is distinctive in that it plays no part in contributing to environmental problems at any scale, while yet deriving its *raison d'être* as an actor from the promotion of solutions to those problems (interestingly ENGOs may not contribute to environmental problems, but they do benefit from their existence indirectly in that the perceived seriousness of environmental problems may be linked to the fund-raising capacity of these actors: see, for example, Wapner, 1996). Although there is great diversity amongst ENGOs, there tends nevertheless to be a general commitment on the part of most ENGOs to promoting solutions premised on democratic environmental management practices at the local scale (Fisher, 1993). To the extent that ENGOs can bring outside funding and media attention to bear on a given environmental problem, they may be able to strengthen the campaigns of grassroots actors.

The preceding discussion has sought to give a sense of how scale, environmental problems and actors are potentially interrelated. The complexity of these interrelationships needs to be emphasised. Table 2.2 has simplified these connections in order to explore them at various scales (and with reference to selected examples only). What it does not show, therefore, is the fact that environmental problems may interact at any given scale (i.e. deforestation, soil erosion) as well as between scales (i.e. deforestation, greenhouse warming) in the process exacerbating the environmental crisis. Further, just as environmental problems may interact at various scales, so too different actors interact simultaneously over the environment at different scales, with the result that developments at one scale may have an important bearing on activities at another level. To give but one example: a state and an ENGO may come into conflict at the global scale (e.g. at an international conference) over the issue of deforestation, but may also interact within the country in question. The relationship at the local scale may be influenced by interaction between the two actors at the global scale, and vice versa (see Chapter 6). The precise nature of such a relationship will depend on diverse political, economic and ecological conditions. However, in seeking to clarify the meaning and significance of the interaction of these and other actors, political ecologists emphasise the role of power in conditioning who benefits, and who loses, from the Third World's environmental crisis.

POWER

We have suggested so far in this chapter that unequal relations between actors are a key factor in understanding patterns of human–environment interaction and the associated environmental problems that, in aggregate, constitute the Third World's environmental crisis. Those unequal relations need to be related, in turn, to the power that each actor possesses in greater or lesser amounts, and which influences the outcome of environmental conflicts in the Third World. Power is, thus, for political ecologists a key concept in efforts to specify the topography of a politicised environment.

A large literature exists on the political, economic and cultural dimensions of power, which is linked notably to the work of Luke (1977), Foucault (1977), Mann (1986), Cox (1987) and Escobar (1995). However, political ecologists have primarily understood this concept in relation to the ability of an actor to control their own interaction with the environment and the interaction of other actors with the environment. It is above all 'the control that one party has over the environment of another party' (R. Adams cited in Bunker, 1985: 14) that has been a preoccupation in Third World political ecology.

Yet to appreciate the role that power plays in conditioning patterns of human–environmental interaction, it is necessary to adopt a more inclusive understanding of power that encompasses material and non-material considerations as well as the apparent fluidity of power itself. The following discussion seeks to do so by addressing three interrelated questions: (1) What are the various ways and forms in which one actor seeks to exert control over the environment of other actors? (2) How do power relations manifest themselves in terms of the physical environment? (3) Why are weaker actors able to resist their more powerful counterparts?

There are various ways in which one actor may seek to exert control over the environment of other actors. First, and most evidently, an actor can attempt to control the access of other actors to a diversity of environmental resources (e.g. land, forests, water, marine/terrestrial wildlife, minerals). The objective here may simply be to monopolise a valued environmental resource so as to ensure that the economic benefits associated with the exploitation of the resource in question accrues largely, if not exclusively, to the actor. A classic example of this strategy is the effort by colonial states in nineteenth- and early twentieth-century Africa and Asia to monopolise control over selected forests containing commercially valuable timber such as teak or cedar (Guha, 1989; Peluso, 1992; Bryant, 1997a).

However, and as this example further illustrates, an actor such as the colonial state also frequently needed to control the access of other actors to those areas in which the valued resource was located. The protection of teak forests in British Burma or Dutch Java involved the creation of a network of reserved forests from which farmers and shifting cultivators were excluded or to which they were permitted restricted access only (Peluso, 1992; Bryant, 1994b; see also Chapter 3). Colonial states (and postcolonial states subsequently) have thus sought to exert control over selected actors and environments in the pursuit of their interests through a policy based on partial or total exclusion. To the extent that they have succeeded (but see

39

below and Chapter 7), states have demonstrated their power over other actors in so far as they have been able to determine who exploits selected environmental resources, the conditions under which those resources are exploited, and often even for what purposes they are used.

This point can be related to an earlier argument of this chapter. Power manifested as control over access is linked to a marginalisation of weaker grassroots actors which also often leaves the latter vulnerable to episodic changes (see Table 2.1). It may also be linked simultaneously to a highly unequal distribution of the costs and benefits associated with emerging environmental problems – viz. the weak/marginalised actors are especially hard hit by the costs, while the more powerful actors are able to capture disproportionately any benefits (Table 2.2). Indeed, the link between such costs (e.g. land degradation) and weaker actors is merely the flip-side of the connection between powerful actors and environmental benefits acquired through privileged access to environmental resources.

An actor may also be said to exert control over the environment of others in so far as it is able to influence or determine the location of the sites at which industrial pollution is generated and released into the environment. Power, here, is about attempting to avoid, or at least minimise, the costs associated with the manufacturing process. On a global scale, this point is illustrated by the gradual shift of selected 'dirty' industries (e.g. chemicals, asbestos) from the First World to the Third World. To be sure, businesses do not make this move based only on calculations as to where environmental regulations are the most lenient – the so-called 'pollution haven' thesis (Pearson, 1987; Leonard, 1988). However, the desperate need of most Third World countries to develop out of poverty, and the perception of most Third World leaders that industrialisation is the only way to achieve such development, has led to a sharp discrepancy between pollution regulatory regimes in the First and Third Worlds. The result has been the concentration of highly dangerous industries in such countries as Thailand, Indonesia, Brazil or India, and growing pollution and health problems there on a scale not seen in the First World since the Industrial Revolution (Hardoy et al., 1992).

Yet as the 1984 Bhopal disaster demonstrated, the power to avoid industrial pollution is not simply an issue between First and Third Worlds. The release of toxic gases from the American-owned Union Carbide plant in Bhopal (India) killed or maimed thousands of mainly poor people living adjacent to the plant (Weir, 1988; Morehouse, 1994). However, the location of dangerous and under-regulated industrial plants primarily in poor neighbourhoods in Third World cities (as in selected First World cities: see Bullard, 1993) is no accident, but reflects the power of selected actors (e.g. states, businesses) to determine the spatial distribution of hazardous activities and, by extension, to determine which groups in society are most likely to be exposed to the potential costs associated with such activities. The ability to control exposure to potentially life-threatening industrial practices is perhaps the most striking, if yet poorly understood, illustration of how one actor is able to exert control over the environments of other actors (Chapter 5).

An actor can also seek to exert control over the environment of others through control over the societal prioritisation of environmental projects and problems. We are thinking here of how actors inside and outside the state can influence the environmental management priorities of state agencies so as to favour the allocation of financial and human resources to certain environmental projects and problems, as opposed to other projects and problems. The point here is not about the power to control environmental resources or the incidence of pollution per se, but rather about the power to allocate the financial resources of the state so as to intervene in different types of human–environment interaction either to 'remedy' diverse environmental problems or to establish/support 'desirable' environmental projects.

The provision of official subsidies for the 'green' activity of plantation forestry is a case in point. As noted above, the role of business in efforts to 'solve' the problem of deforestation through plantation forestry has grown considerably in recent years. The influence of businesses involved in this sector over state leaders and officials in such countries as Brazil, Thailand or Indonesia is now considerable (Puntasen et al., 1992; Marchak, 1995; Lohmann, 1996). Such influence is reflected not only in a favourable regulatory regime, but also in the provision of selected state subsidies to promote this form of 'sustainable' forestry. However, plantation forestry is often opposed by poor farmers and other grassroots actors who, acting frequently with the support of ENGOs, urge the state to support alternative community forestry projects. The plantation/community forestry struggle is but one example of how different actors assert competing claims with the aim of influencing the state in its prioritisation of environmental problems and issues – and how this struggle is linked to efforts by selected actors to control the environments of others.

Finally, an actor may seek to exert control over the environment of other actors in an indirect manner through discursive means. Power is about control over material practices, but it is also linked to the attempted regulation of ideas. That ideas are never 'innocent' but 'either reinforce or challenge existing social and economic arrangements' (Schmink and Wood, 1987: 51) was noted in Chapter 1. The implications of this statement for our understanding of the role of power in Third World environmental conflicts is twofold. First, it serves to emphasise that power is partly a matter of 'winning the battle of ideas' over human use of the environment, since actors typically seek to legitimate the triumph of their individual interests over the interests of others through an attempt to assimilate them to 'the common good' (Schmink and Wood, 1992). To take the forestry example again, states have not been content merely to exert physical control over designated forests at the expense of other actors, but have also sought to justify this move in terms of the 'ecologically bad' practices of the latter (i.e. the 'destructive' shifting cultivator) compared with the 'ecologically good' practices of the state. In effect, states and other powerful actors seek to maintain or enhance their power over the environments of other actors by controlling what J. Scott (1990) terms the 'public transcript' – that is, the 'socially accepted' version of events represented in public documents, legal political ideologies, popular music and theatre, and so on. Through control of the public transcript, actors seek to render 'natural' the triumph of their partisan interests on a society-wide basis (see also Peet and Watts, 1996b; Escobar, 1996).

Yet the statement by Schmink and Wood also hints at the considerable difficulties facing powerful actors when they seek to do so. Distopian visions of thought control notwithstanding (e.g. Orwell's novel *1984*), the realm of ideas remains distinct from the realm of material practices – the former cannot simply be 'read off' from the latter (and vice versa). As Schmink and Wood (1992: 16–17) note in this context, 'ideological positions ... are not mere reflections of material interests. Nor are they static features of people's consciousness ... [rather] they are mutable and subject to continual redefinition.' However, it is precisely because the realm of ideas cannot be seen, and hence cannot be rigorously 'policed', that powerful actors often find it so disturbing. Yet as J. Scott (1990) shows, 'hidden transcripts' need not be confined to the imagination of an individual actor but can serve as the basis for a 'culture of resistance' (Peluso, 1992) among oppressed actors. This point will be returned to again below when the dangers associated with a reading of power based exclusively on material practices are noted.

Another way in which to understand the role of power in conditioning human–environmental interaction is to consider the physical environment as a manifestation of power relations. The idea here is not to focus on the specific material and discursive interactions of actors, but rather on the environmental 'results' of those interactions. In other words, what can we learn about power relations, past and present, from a careful 'reading' of the environmental 'text' in the Third World? Much work has already been undertaken, notably in environmental history (e.g. Worster, 1985), cultural geography (e.g. Cosgrove and Daniels, 1988) and environmental sociology (e.g. Greider and Garkovich, 1994), which explores the 'iconography of landscape' (cf. Moore, 1996). Even more ambitious has been the attempt to quantify power in terms of control over energy flows (Adams, 1975) – a theme, moreover, explored in at least one key political-ecology text of the 1980s (Bunker, 1985). However, the latter illustrates nicely the immense difficulties associated with any attempt to quantify the material aspects of power through the measurement of energy flows – let alone the question of how one is to interpret the discursive aspects of power.

A reading of the environmental text is nonetheless useful in that it can cast further light on how unequal power relations among actors are 'inscribed' in the environment. What are some of the elements of such a reading, and what do they tell us about power and the Third World's politicised environment? To begin with, it is possible to 'visualise' unequal power relations in so far as it is possible to specify spatial patterns of control and resistance. Thus, patterns of control involve powerful actors shaping the use of environmental resources through such economic activities as logging, mining, manufacturing or cash cropping. These activities are inscribed, in turn, in the environment. Felled forests, timber plantations, cotton fields, toxic waste dumps, human-induced concentrations of big game, tailings around mines, or air and water pollution near factories, may all reflect the 'imprint' of powerful actors (e.g. states, businesses). At a more general level, the specification of national boundaries itself reflects an attempt by the state to map which actors and environments fall under its control (Walker, 1993; Vandergeest and Peluso, 1995).

In contrast, patterns of resistance are often more difficult to discern in the environment precisely because, as Scott (1985) notes, the weak rarely seek to draw public attention to their resistance. However, patterns of resistance do occur, and are notably associated with the 'illegal' exploitation of environmental resources by poor farmers and other grassroots actors. Forest clearances linked to illegal cultivation or fuelwood gathering in national parks or reserved forests, the poaching of big game in wildlife parks or the cultivation of forbidden crops (e.g. coca), may all be examples of these actors asserting their perceived right to shape local environmental conditions. Similarly, efforts by these actors (often acting in conjunction with ENGOs) to specify the extent of local community forests or 'ancestral domain' represent an attempt to 'counter-map' contested environments (Colchester, 1993; Walpole et al., 1993; Peluso, 1995).

In addition, specific nodes of control and resistance can also be highlighted to clarify how unequal power relations are reflected at key points in the physical environment. Thus, nodes of control can be conceived of as infrastructural projects that facilitate wealth creation and the maintenance of political 'order': large hydro-electric dams (energy), industrial plants (manufactured goods) or roads (security) all serve to reinforce, and in some cases even symbolise, the power of stronger actors (Allen, 1992; Usher, 1996). In contrast, the identification of nodes of resistance can be a more difficult task, yet is to be understood in terms of infrastructure that potentially serves as the basis for autonomous local community livelihoods and social organisation. The main example here is the small-scale irrigation networks found throughout much of the Third World (e.g. *zanjeros* in the Philippines, *mu'ang fai* in Thailand) which have often been a long-standing source of power for small-scale farmers in relation to external actors (Ostrom, 1990; Rigg and Stott, 1996).

It is thus possible to 'read' the environment to get some sense of how unequal power relations are reflected in the environment. Yet, as suggested above, such a reading is insufficient on its own because it fails to capture the many 'intangible' qualities associated with the concept of power. One way in which to appreciate these qualities is to consider why it is that weaker actors operating in contexts of often highly unequal power relations are able to retain any power at all.

The ability of weaker actors to resist their more powerful counterparts is linked to the great difficulties that even the most powerful of actors face when seeking to exert control over the environments of others. Most grassroots actors relate to the environment in complex ways reflecting a multiplicity of material and cultural interests (Blaikie and Brookfield, 1987), as well as the vicissitudes of local ecological conditions (Zimmerer, 1994). However, powerful actors often seek to impose a new political and ecological order, thereby overriding existing local patterns of human–environmental interaction. In some cases, a short-term strategy is followed whereby these actors exploit local resources until they are depleted, and then move on to a different locality to repeat the process. The destruction of most of South-East Asia's forests as a result of large-scale logging by states and businesses is a case in point (Hurst, 1990; Bryant *et al.*, 1993). Under these circumstances, the ability of weaker grassroots actors to resist can be quite limited.

In other cases, a more complex situation ensues as powerful actors seek to build up a long-term stake in a given locality, but in the process provide an opportunity for weaker actors to resist development practices that they perceive to be inimical to their interests. To return to an example used earlier in this chapter, states and businesses have devoted considerable financial and human resources to the creation of commercial forest plantations (e.g. teak, eucalyptus, pine) designed to tap the growing worldwide demand for timber as well as pulp and paper (Marchak, 1995). Yet, as colonial officials discovered in the nineteenth century and as their postcolonial successors are still finding today, these plantations are an inviting target for disgruntled grassroots actors whose interests are prejudicially affected by their creation (Guha, 1989; Peluso, 1992). Since the nineteenth century the spread of commercial forest plantations has been associated with arson and other acts of sabotage as a form of 'everyday resistance' by grassroots actors (Scott, 1985). States and businesses have experienced great difficulty in countering these weapons of the weak (see Chapter 7). Powerful actors thus seek to control human–environment interaction in the pursuit of their interests, but in doing so often leave themselves open to attack by weaker actors. The ensuing struggles may be highly uneven, but the ability of the latter to use their local environmental knowledge to subvert the activities of the former is not to be gainsaid.

A further reason why weaker actors may be able to resist their stronger counterparts relates to the question of legitimacy. Powerful actors typically feel the need to justify their control over the environments of others in terms of appeals to the 'common good' as part of their attempt to 'naturalise' the situation (see above). The state as a public institution is especially keen to seek legitimacy for its actions. Indeed, the whole notion of the state as a protector or 'steward' of the environment is bound up with the perceived need for popular legitimation (see Chapter 3). Yet powerful non-state actors also seek to justify their actions on grounds other than self-interest. Powerful farmers have long explained their privileged social and economic position partly in terms of an 'obligation' to protect the interests of marginal members of the community, especially during times of dearth (Scott, 1976). Transnational corporations habitually justify their activities in terms of the employment and other benefits (i.e. taxes) that accrue to communities as a result of their presence (Pearson, 1987).

As noted above, powerful actors are keen to shape the 'public transcript' that surrounds the question of legitimacy. Yet, the alternative 'hidden transcripts' of weaker actors always pose a danger for powerful actors precisely because they question 'official history'. And, on those occasions when the hidden transcript becomes public – that is, when weaker actors openly challenge the claims to legitimacy of more powerful actors – the political and economic situation becomes explosive (J. Scott, 1990). Although the more powerful actors may well prevail in the ensuing struggle, their power is nonetheless diminished in so far as the activities that are the source of their power are seen as 'illegitimate' by the wider population.

The ability of weaker actors to resist stronger actors also derives from the fact that it is very rarely the case that one actor possesses overwhelming power over all other actors. As a result, in the relatively fluid circumstances that are associated with multiple power centres, weaker actors are often well placed to assert their interests vis-à-vis more powerful actors. It is the state which has come closest to possessing overwhelming power in relation to other actors, especially in the former socialist countries (Smil, 1984; Beresford and Fraser, 1992; Hershkovitz, 1993). Yet to hold a formal monopoly on the means of coercion in society has rarely meant that the state has been in a position simply to impose its interests at will. Rather, state actions are conditioned by the relationship of the state to other actors both inside and outside the country. It is true that coalitions of the powerful – notably states, businesses and multilateral institutions – have frequently exacerbated the plight of weaker actors (Adams, 1991; Rich, 1994). Yet those coalitions are rarely without internal divisions, and the state itself is often subject to powerful fissiparous tendencies as a result of bureaucratic conflict (see Chapter 3). The growing power of ENGOs has altered the power equation further, typically to the detriment of traditionally powerful actors. ENGOs put the spotlight on the activities of states, businesses and multilateral institutions thereby rendering it difficult for these actors to control the public transcript (Princen and Finger, 1994). Indeed, the key contribution of ENGOs in contemporary political-ecological conflicts is arguably to publicise and support the struggles of poor farmers or shifting cultivators (Broad, 1993; Eldridge, 1995). The ability of traditionally powerful actors to control weaker actors has been slowly but surely eroded in the process (see Chapters 6 and 7).

A final reason why weaker actors are potentially able to resist stronger actors relates to the tools of power. To some extent, the power of a given actor is a function of technology – simply put, more powerful actors develop, control and disproportionately benefit from technological change (Hill, 1988). For example, the colonial state derived power over its subjects in part through use of the 'tools of empire': quinine, the machine gun, the steamboat, the railway and the telegraph (Headrick, 1981). Powerful state and non-state actors today continue to benefit disproportionately from technological changes linked, in particular, to the new computer or biological technologies (e.g. Kloppenburg, 1988). Yet, as Mann (1986) points out, actors are rarely able to monopolise for long technological innovations or their associated political and economic benefits. To take two contemporary examples, the facsimile machine and the Internet may be useful to state agencies and businesses in the pursuit of their political or economic interests, but these technologies are also used by less powerful groups, sometimes even to mount campaigns attacking the practices of powerful actors.

In a similar fashion, the power long associated with organisational techniques has become more diffuse in as much as new actors develop comparable capabilities. Thus, the power of the state has derived in part from its institutional complexity (especially since the nineteenth century), which has permitted it to coordinate and control often quite heterogeneous peoples and environments. Critical to this endeavour has been the state's use of statistics and maps to specify precisely the people and environments under its control (Anderson, 1991). Yet transnational corporations, multilateral institutions and ENGOs are institutionally complex today, and also use statistics and maps to assert their own interests (Rich, 1994; Princen and Finger, 1994; Welford, 1996). The initial organisational advantage enjoyed by the state has thereby been eroded, and with it, often the power of that actor in relation to other actors.

The preceding discussion has emphasised the complexity of power relations. It has suggested that the ability of an actor to control or resist other actors is never permanent or fixed but always in flux. Thus, power influences the topography of a politicised environment, but in such a manner that the relative position of actors can never be adduced exclusively from material considerations. It was the tendency of some early political ecology work to do so that prompted the charges of economic determinism and apoliticism discussed in Chapter 1.

To understand the workings of a politicised environment is to appreciate, therefore, the complex ways in which actors interact at the material and discursive levels over environmental questions. Yet the suggestion that power is at the heart of this politicised environment serves as a reminder that it is the unequal power relations of *actors* that is central to research in Third World political ecology. The aim of the remainder of this book is thus to explore the role of selected key actors in environmental management and conflict so as to evaluate the distinctive political contribution of each type of actor to human–environmental interaction in the Third World. As the next chapter shows, the state has played an especially prominent role in such interaction.

3

THE STATE

To appreciate the ways in which environmental change is politicised in the Third World is in considerable measure to understand how the state has sought to manage the peoples and environments within its jurisdiction. Chapter 2 noted that most actors contribute to environmental change as they pursue their interests, but it is the state which has traditionally played a leading role in conditioning how diverse actors interact with the environment. To speak of a politicised environment is thus to signal the historical and contemporary importance of this actor.

This chapter explores the role of the state in environmental conflict and management in the Third World. It is often said that the power of the state derives largely from this actor's unique remit to act in the 'national interest'. Yet if society's need for an institution such as the state is well established in theory, the practical dilemmas facing all Third World states in terms of integrating the concerns of economic development and environmental conservation are legion. The chapter also considers the possible contradictions in the state's dual role of developer and protector of the natural environment, and how such contradictions may be reflected in terms of intra-state and inter-state conflict. The question as to whether the rise of other actors, notably transnational corporations and environmental non-governmental organisations (ENGOs), signals a decline in the state's hitherto privileged role in environmental matters is then taken up, and provides a fitting introduction to themes explored further in subsequent chapters.

THEORY INTO PRACTICE

The origin of the state is attributed in the theoretical literature to the perceived failure of individuals to behave in a socially responsible manner. People pursuing individual interests 'inevitably' become embroiled in conflict with one another in the absence of a state capable of imposing order in the collective interest. Yet if this theoretical argument explains why the state is needed, it does not explain how it acquired power in practice.

The theoretical justification for the state revolves around a linked set of assumptions concerning human behaviour, collective interests and the capabilities of the state to pursue such interests. The best known early statement on this subject was provided by the British philosopher Thomas Hobbes in 1651 who described how, in a stateless world, people were required to pursue their individual interests in a context of social anarchy. This was so because in the absence of plentiful environmental and social resources people were forced to acquire power over the labour and environments of others so as to achieve social and economic wellbeing. However, since each person sought power over others, a highly destructive process ensued in which extreme social and economic uncertainty were the norm: the 'life of man' under these conditions could only be 'solitary, poore, nasty, brutish, and short' (Hobbes, 1968). To avoid such a fate, Hobbes argued, rational individuals would recognise the need for a state to impose order on people, and they would voluntarily surrender some of their liberty in order to allow the state to do so. The price of order was thus a sovereign state with a monopoly on the means of coercion within a given territory.

The theoretical basis for the state has been elaborated since Hobbes's time, but the underlying assumption that individual action in the absence of a state can lead only to anarchy remains largely unchanged. Thus, for example, game theory has explored the obstacles preventing the development of trust and cooperative behaviour between 'rational' individuals, notably with reference to the Prisoner's Dilemma game (Axelrod, 1984; Johnston, 1989). This game uses the hypothetical example of two men arrested by the police for committing a minor offence, but who are suspected of being responsible for a serious crime. The police interview the two suspects separately and, to acquire conclusive evidence, present each man with a deal: 'squeal on your accomplice, and if he is convicted of the major crime, go free; or stay silent, and be convicted of the lesser offense and go to jail'. Figure 3.1 illustrates the choices facing each man.

The collective interest of both men is to stay silent since that option provides the lowest aggregate sentence. However, it is in each man's individual interest to squeal, provided that the other man does not follow suit, because that way he is able to escape any punishment. Both men separately reach this conclusion and squeal accordingly with the inevitable result that they are both then convicted of the major crime. This result is the worst possible collective solution and illustrates the main lesson of Prisoner's Dilemma – namely that in the absence of trust (or a state to enforce 'trust') between individuals, it is inevitable for people to act individually, but with an outcome that is sub-optimal from the viewpoint of society as a whole.

The case for a collective interest articulated and enforced by the state is perhaps nowhere more vividly described than in Hardin's (1968) essay on 'the tragedy of the commons'. Hardin describes a situation in which individual herders graze their cattle on a common pasture. Each herder seeks

Prisoner B

		Stay silent	Squeal
Prisoner A	Stay silent	1,1	10,0
	Squeal	0,10	8,8

Figure 3.1 Prisoner's dilemma
Source: Johnston, 1989: 114

to increase the number of cattle on the commons until eventually the 'carrying capacity' of the land is exceeded. However, rather than curtailing use of the commons, each herder continues to add cattle to the pasture, resulting inevitably in tragedy as the land is degraded and the livelihoods of the herders are ultimately impoverished. The 'inherent logic' to this process derives from the fact that the benefit associated with each additional animal accrues entirely to individual herders whereas the costs of additional cattle accrue to all herders. As with Prisoner's Dilemma, individual action in Hardin's metaphor leads inexorably to social and environmental ruin in the absence of a state to protect collective interests.

Using theoretical arguments based in part on Prisoner's Dilemma and the tragedy of the commons, various scholars have argued that there is a need for an omnipotent state to tackle the world's growing social and environmental crises. This book has already touched on the neo-Malthusian school in terms of its likely impact on the early development of Third World political ecology. Chapter 1 suggested that early work in the latter may have shied away from using the political ecology label during much of the 1970s as a result of the association of that label with a neo-Malthusianism that most radical scholars considered abhorrent. Here, it is useful to note that the neo-Malthusian school has also been described in the literature as 'neo-Hobbesian', in recognition of its central assertion of the need for a global Leviathan. Various writers contributed to this argument (e.g. Heilbroner,

1974; Hardin and Baden, 1977), but it received its most sophisticated treatment in the work of Ophuls who argued that

> ecological scarcity in particular seems to engender overwhelming pressures toward political systems that are frankly authoritarian by current standards, for there seems to be no other way to check competitive overexploitation of resources and to assure competent direction of a complex society's affairs in accord with steady-state imperatives. Leviathan may be mitigated, but not evaded.
>
> (Ophuls 1977: 163)

Thus, according to Ophuls, social complexity and ecological crisis rendered authoritarian state action seemingly inevitable.

Much of the criticism of the neo-Hobbesian argument centred on the ways in which it exalted the position of the state in relation to other actors. The authoritarian implications of what Enzensberger (1974: 11) terms 'environmental protection from above' were condemned vigorously as being both unworkable and ignorant of basic considerations of social justice (Orr and Hill, 1978; Hoffert, 1986; Hayward, 1995). Some scholars also criticised the neo-Hobbesian interpretation of the work of Hobbes itself (Walker, 1988), while other writers sought to show that Hardin's tragedy of the commons metaphor lacked explanatory power in 'real-world' situations (McCay and Acheson, 1987). These criticisms cast doubt on the validity of the neo-Hobbesian argument but also raise much broader questions about the theoretical justification for the state itself.

These questions involve a reassessment of what has been seen hitherto to be a theoretical strength of the state – namely, its role as a sovereign actor holding a monopoly on the means of coercion within a given territory. The empirical conditions that have prompted this reassessment are considered below, but here the concern is to set out the theoretical aspects of that critique.

One aspect of the theoretical critique of the state concentrates on the possibility that this actor represents an obstacle to the resolution of environmental problems at the global scale. The argument here goes as follows. The primary goal of the world's states has been to pursue economic development even if, as experience has shown, this quest has been at the expense of the environment (Walker, 1989; and see below). The almost inevitable result is the growth of environmental problems at the local and regional scales, but also increasingly at the global scale (see Chapter 2). Yet little remedial action is taken as states guard their sovereignty jealously against proposals for a globalised system of environmental management (Mische, 1989). Concurrently, few states are prepared to eschew the benefits of economic development in the context of an increasingly competitive global capitalist economy, especially when the resulting environmental costs in terms of various forms of pollution are often distributed globally.

Ironically, a tragedy of the global commons ensues in which individual states (i.e. herders) continue with policies and practices that degrade the global environment (commons), while refusing all the while to give up the individual right to action (sovereignty) that is at the root of the problem (Vogler, 1995).

A second and related aspect to the theoretical critique of the state focuses on the possible incapacity of the state to address effectively environmental problems at any scale. The argument here is that the state as an actor is either 'too small' to manage regional and global environmental problems or 'too big' to deal with local environmental problems (Hurrell, 1994). This argument suggests that there may be an inherent discrepancy between the essential traits of the state as an actor and the social responses needed to address environmental problems that rarely, if ever, respect national boundaries (Turner *et al.*, 1990). Thus, states have been described as 'Janus-faced' actors that derive much of their power from being at the nexus of the national and international political orders (Mann, 1984). Yet there is little evidence to indicate that this source of social power is even remotely compatible with sustainable environmental management. Indeed, the weight of evidence suggests the contrary view. Thus, on the one hand, states have tended to hinder the environmental initiatives of grassroots actors at the local scale while, on the other hand, preventing efforts to develop a comprehensive global approach to solving the world's global environmental problems.

These critiques of the theoretical basis of the state are perhaps best understood in relation to the empirical developments that have conditioned the possibilities of state power over time (see below). Yet if the preceding discussion has shown anything at all, it is that the role of the state is contested as fiercely at the theoretical level as that role has been challenged at the empirical level.

To appreciate why that role has been so fervently contested over time, it is necessary to understand the rise of the state as a leading social and environmental actor since the seventeenth century, if not before. The historical development of states has been closely intertwined with the management of the local environments on which those states, and the people they govern, have been dependent. The need to extract an economic surplus in order to maintain or increase state power was associated with a relentless quest to maximise natural resource production, often beyond 'sustainable' levels, even in ancient times (Walker, 1989). Deforestation, for example, has been linked to the policies and practices of early states in the Mediterranean basin and China (Thirgood, 1981; Smil, 1984). Yet it was the rise of the modern state in Europe beginning in the seventeenth century, if not before, which marked the ascendancy of this actor as the leading player in human–environmental interaction (Hall, 1986).

Many scholars have linked the growing power of the modern state to the development of global capitalism (e.g. Wallerstein, 1974; Wolf, 1982;

Johnston, 1989; Blaut, 1993). The general argument is that an actor such as the state was required to provide diverse public goods ranging from security and a common currency to social and physical infrastructure (i.e. roads, education), and that without such intervention the capitalist system would not have prospered because it would have been unable to accumulate capital (the key activity in the capitalist system). As Johnston (1989: 70) notes, states 'must be there to do certain things, otherwise capitalism will fail'. The state is thus an institutional necessity under the capitalist system because without its presence a Hobbesian anarchy would prevent capital accumulation. From this perspective, the rise of the state can be seen to be a case of an institution being in the 'right place at the right time'. It is for this reason that the state is seen by many scholars (including political ecologists: for example, Watts, 1983a; O'Brien, 1985) to be inextricably linked to the advance of capitalism as a mode of production.

However, that the modern state is closely associated with capitalist development is not to say that it does not have distinctive interests or sources of power. Indeed, the state often has its own political, economic and strategic interests that derive from its unique socio-spatial position at the intersection of the domestic political order and the inter-state system (Skocpol, 1985; Mann, 1986; and see above). What this means in practice is that the interests of the state and capitalists do not always coincide. A case in point is state efforts to promote selective environmental conservation in the face of business opposition. Often, conflict has centred not on whether an environmental resource is to be exploited commercially, but rather on the conditions under which such exploitation is to occur – viz. restrictions on logging in order to ensure long-term production versus the triumph of *laissez-faire* 'cut-and-run' practices (Guha, 1989; Bryant, 1994b).

The ability of the state to enforce its will in the face of opposition from business or other actors has been the result in part of its capacity to take advantage of a succession of technological innovations that enhanced the state's coercive and surveillance powers (see also Chapter 2). Thus, advances in military technology facilitated the imposition of state control over hostile populations and in peripheral areas. In the nineteenth century, for example, the European colonial states used such inventions as the machine gun and the steamboat to assert military control over much of what later became known as the Third World (Headrick, 1981). Concurrently, advances associated with the advent of the railways or the telegraph permitted states to link together disparate and hitherto isolated territories. The building of physical infrastructure (including road and canal networks) served to knit together territories, thereby facilitating central state control. Such control was further enhanced through the accumulation of detailed knowledge about the location of people (along with their sources of livelihood) and environmental resources through the use of maps, surveys and statistics (Anderson, 1991).

The development of the modern state was closely linked to the European conquest of much of the Third World, notably in the nineteenth century (Watts, 1983a; Blaikie and Brookfield, 1987). On the one hand, European states used the means noted above to seize control of Third World countries, in most cases overpowering indigenous states in the process. On the other hand, the advent of colonial rule itself was a powerful fillip to state empowerment, as it triggered a flow of environmental resources from the Third World to the First World, which enriched European state coffers, permitting in turn a rapid expansion in the size and scope of the colonial state (Taylor, 1987). Notwithstanding the pleas of Adam Smith and other classical economic thinkers for a minimal state, the nineteenth century bore witness to the growth and functional diversification of the state in both 'home' and 'colonial' territories.

A related development was the quest of colonial states to base their administration of conquered peoples and environments on 'scientific' principles in order to 'rationalise' the social and natural environments under their jurisdiction (Richards, 1985; Adas, 1989). Yet such social and environmental 'engineering' was met by widespread popular resistance in the nineteenth and twentieth centuries throughout the Third World. To take but one example, the attempt by colonial forest departments in South and South-East Asia to reorganise the forested landscape through the segregation of imperial-commercial and local (i.e. 'minor forest products') use areas was met by concerted 'everyday' resistance as grassroots actors (often acting in cahoots with indigenous businesses) extracted forest products illegally from state forest reserves (Guha, 1989; Peluso, 1992; Bryant, 1994b). As Chapter 7 shows, the result was a process of environmental conflict, founded on a dynamic of attempted state control and grassroots resistance, that has persisted in many parts of the Third World to the present day.

The preceding discussion has been a broad-brush explanation of the theoretical and historical significance of the state as a key actor in problems of environmental management and conflict. It has highlighted, among other things, the existence of a central tension in the 'mission' of the state as an actor linked to its combined role as developer and steward of the environment, a theme to which we now turn in greater detail.

TO DEVELOP OR DESTROY THE ENVIRONMENT?

The subject of the state and environmental management has evoked much pessimism amongst scholars. Johnston (1989) argues that the state's role as the facilitator of the capitalist system links that actor to contemporary environmental problems that are an essential byproduct of that system. Yet the state is also the key actor involved in finding a solution to those environmental problems, but it is largely prevented from doing so because it is more or less beholden to the interests of capitalists. Walker (1989: 32) gloomily adds that 'explicit state responsibility for management of the biological and physical resource base, though effectively unavoidable, has never been accepted'. Such pessimism stands in sharp contrast to the expectations of state behaviour derived by some writers from theory (see above). After all, the state is an actor that is supposed to be dedicated to the promotion of collective goods, of which the environment is a leading example. Thus, while the historical rise of the state might seem to confirm theoretical expectations about its general necessity, in practice state behaviour – inasmuch as the environment is concerned – has been disappointing, if not disastrous. Rather than being an actor with possible solutions to environmental problems, the state has typically contributed to exacerbating those problems.

At the heart of any explanation of why states have been so destructive environmentally must be the recognition of a central paradox in the state's function. In effect, there is 'an inherent, continuing potential for conflict between the state's roles as developer, and as protector and steward of the natural environment on which its existence ultimately depends' (Walker, 1989: 32). This paradox will be assessed through a consideration of the environmental management practices of the state in the Third World since the Second World War (when many countries in Asia and Africa gained independence).

Environmental conservation was a low priority immediately following the Second World War for most states in Africa, Asia and Latin America. The immediate goal was the assertion of political control by the state over the people living in the territory under its jurisdiction, often in a context of widespread social unrest. As a result, environmental policies developed in colonial times were often abandoned even as the management structures associated with those policies frequently remained in place. Forest policies carried over from colonial times in Burma and Indonesia, for example, were largely inoperative as a result of pervasive civil unrest, even though the colonial forest services in these countries persisted. Thus, the onset of civil war in Burma following independence in January 1948 resulted in a situation in which Burmese forest officials required armed escorts to extract teakwood from the main Pegu Yoma forests well into the 1970s. However, such 'forestry-on-the-run' did not permit the planting and weeding operations that were essential to the colonial-style scientific forestry policies to which these officials swore allegiance (Bryant, 1997a). A similar situation prevailed in Indonesia after independence was won from the Dutch in 1949. Forest use and policy was politicised so that Dutch-inspired forestry management programmes could not operate successfully. Struggles over the forest were 'localized versions of larger struggles for power at the national level' in a process that reached its bloody climax in the events of 1965–6 when thousands of communist members or sympathisers were either imprisoned or lost their lives in the army crackdown led by President

Suharto (Peluso, 1992: 122). Civil unrest of this kind only encouraged political leaders to ignore the state's own environmental policies as part of a quest for political survival. Faced with the prospect of political or military defeat, few leaders had any qualms about encouraging 'cut-and-run' resource extraction strategies in areas controlled by the state in order to maximise short-term revenue. Environmental conservation under these conditions was merely an obstacle to the survival of the political regime itself.

The link between political security and the neglect of the state's stewardship role was not confined exclusively to the colonial transition in Africa and Asia. In Latin America, concerns about national security in the context of boundary disputes and communist agitation led to policies specifically designed to transform the environment as part of a nation-building exercise. In Brazil, for instance, successive governments have viewed the vast Amazon region as an area desperately in need of 'development'. As political ecologists have amply documented, the encouragement of cattle ranching, mining and other commercial activities has resulted in large-scale deforestation and associated environmental degradation (Hecht and Cockburn, 1989; Schmink and Wood, 1992). The pressure for such development came, at least in part, from a Brazilian military keen to secure remote northern and western borders against rival neighbouring states (Allen, 1992).

An unstinting quest for economic development in the Third World after 1945 only added to the pressures on the state to neglect its stewardship role. Most states drew up ambitious development plans (often with the help of Western economic advisors) in which rapid industrialisation was central to the attempt to catch up with the First World (Peet, 1991). Expensive capital goods (e.g. equipment) were required to initiate industrial development and, to pay for these goods, Third World states had little choice but to maximise natural resource exports in order to offset soaring import bills (Elliot, 1994). Concerns about environmental conservation were predictably absent from most official development plans during the 1950s and 1960s.

Socialist states were no exception to this rule. What is striking when comparing the state's environmental management role in capitalist and socialist Third World countries after 1945 is not the differences, but rather the similarities, between states operating under widely differing political ideologies. Thus, under both ideologies, the quest to industrialise was associated with pervasive resource depletion and environmental degradation. Indeed, as the experiences of China (Smil, 1984; Hershkovitz, 1993) and Vietnam (Beresford and Fraser, 1992) illustrate, the fact that the socialist state typically assumed a bigger role in shaping social and environmental change than its capitalist counterpart meant that, if anything, environmental degradation was worse in the former group of countries than in the latter. The 'transition' to a more market-based approach in the former socialist states in recent years has nonetheless often been associated with a further deterioration in environmental quality (Bryant, 1997a; Muldavin, 1996).

The environmental implications of the Third World's fifty-year-long quest to industrialise have been twofold. First, this state-sponsored quest has been associated with efforts to maximise natural resource extraction as Third World states have emphasised the extraction for export (mainly to the First World) of timber, minerals, fish and cash crops. This process has had the effect of accelerating the political and economic marginalisation of weaker grassroots actors which had often begun under colonial rule. For example, as forests were felled for timber in South-East Asia (Hurst, 1990; Dauvergne, 1993/4), or prime lands were dedicated to groundnut or cotton production in sub-Saharan Africa (Franke and Chasin, 1980; Watts, 1983a), poor farmers or shifting cultivators were usually displaced from their lands to make way for 'modern production' techniques. Expanded commercial food and fibre production was also used to meet rapidly growing domestic needs, notably associated with the burgeoning urban areas (Rush, 1991; Shiva, 1991a). The foreign exchange thereby saved or earned may have enabled the purchase of essential capital goods, but as natural resources were rapidly depleted, Third World states faced a growing environmental crisis within their borders which was partly of their own making.

Second, state-sponsored industrial development has been linked to air, land and water pollution arising from the manufacturing process itself. Third World states have invested a considerable effort both in seeking to attract transnational corporations and in establishing indigenous industry. It is thus not surprising that most states have shown little inclination to regulate in anything other than a perfunctory manner the emissions of industrial plants within their territories. Indeed, the incentive to do nothing in this regard increased after the 1960s, for, as the First World imposed stringent industrial regulations (often as a result of popular protest), the relative lack of pollution controls in the Third World proved increasingly attractive to transnational corporations (TNCs). In conjunction with other considerations such as cheap labour or market proximity, this factor persuaded a growing number of TNCs to re-locate their operations to the Third World in the 1960s and 1970s (see Chapter 5). The development of severe pollution problems in many Third World cities ensued. In Cubatao (Brazil), for instance, rapid and unchecked industrialisation led to intense air, land and water pollution that contributed to abnormally high levels of bronchial and other diseases (Hardoy et al., 1992). The role of these cities as 'pollution havens' (Leonard, 1988) has been contingent in part on the host state, once again, promoting policies that privilege economic development over environmental conservation.

It was becoming increasingly difficult by the mid-1980s, however, for states in the Third World to ignore the growing environmental crisis shaping up within their borders. On the one hand, they were under pressure from First World states to strike a 'balance' between economic development and environmental conservation. Such pressure was associated ultimately with popular environmental concern in the First World which, by the mid-1980s, had grown to encompass 'global' issues, such as tropical deforestation, that, in practice, related primarily to the Third World (see Chapter 6). First World concern about the Third World's environmental woes certainly predates this upsurge of interest in the 1980s and 1990s, and has been reflected both in the policies of United Nations institutions subject to First World control and in the proceedings of environmental summits since 1972 (see below and Chapter 4). Nonetheless, and spurred on by the global campaigns of First World ENGOs such as Greenpeace or Conservation International, First World states have become increasingly critical of their Third World counterparts who perpetuate environmentally destructive policies. They have even resorted to 'green conditionality' – the use of 'environmental goals to condition the objectives, direction and circumstances of aid flows' – so as to attempt to exert influence on the policy process in many Third World countries (Davies, 1992: 151).

On the other hand, Third World states have also been subject to growing pressure for change from diverse actors operating within their own borders. The growing political prominence of grassroots actors has been especially noticeable in this regard, and has been associated with demands by farmers' movements and indigenous people's organisations for a whole new set of policies predicated on social justice, local empowerment and environmental conservation (Friedmann, 1992; and Chapter 7). These actors have received considerable support from foreign and domestic professional ENGOs who have also lobbied states directly to alter traditional policies and practices (Fisher, 1993; see Chapter 6).

The growing environmental 'attentiveness' of Third World states nonetheless merits careful scrutiny against the backdrop of a long-standing commitment to environmentally destructive policies. Beyond the need to distinguish rhetoric from reality, there is also the question of evaluating the political and ecological impact of the 'greening' of state policies in the Third World. It is far from clear in many cases that the flurry of state activity on environmental matters since the mid-1980s constitutes anything other than 'business-as-usual' (Adams, 1993). The specific ways in which Third World states have contributed to environmental degradation through policy incentives is now well documented in the literature. Examples of so-called 'policy failure' (Gupta et al., 1995) include the use of low or non-existent land rental fees, royalties and corporate income taxes, as well as subsidised loans to promote unsustainable logging, cattle ranching or agriculture (Repetto and Gillis, 1988; Hecht and Cockburn, 1989; Hurst, 1990). Yet, notwithstanding growing evidence of the considerable social and environmental costs associated with such activities, most states persist with policies that privilege economic development over environmental conservation.

In Indonesia, for example, the Suharto regime persists with a forestry policy that allows unsustainable logging practices to flourish. The policy is based on legislation passed in 1967, which signalled Indonesia's adoption of an outward-oriented and pro-capitalist economic development strategy. Under this strategy, timber exports grew rapidly in the late 1960s and 1970s as TNCs acting in conjunction with Indonesian partners exploited the country's forested Outer Islands (see also Chapter 5). A ban on raw log exports was introduced by the Indonesian state in the 1980s, but this move was designed to boost plywood production rather than to address mounting environmental problems linked to over-harvesting. Indeed, the ban had the effect of 'locking in' unsustainable extraction levels since the desire to maintain profit and employment levels was enhanced, thereby creating a powerful vested interest in continued maximum production from the forests (Hurst, 1990; Dauvergne, 1993/4). Similarly, in Brazil, pressure to introduce 'green' policies did not stop state support for unsustainable large-scale projects. The economically important but environmentally destructive Grande Carajas programme in eastern Amazonia is a case in point (Hall, 1989; Anderson, 1990). This programme was officially launched in 1980 as 'the most ambitious development programme ever devised for any area of tropical rainforest in the world' (Hall, 1989: xix), and was designed to open up this hitherto remote region through the development of a series of open-cast mines, processing plants and related infrastructure projects (e.g. hydro-electric dams, railroads, a deep sea port). Popular protests certainly prompted minor adjustments to the programme (e.g. limited reforestation); 'nevertheless, large areas of tropical forest were reduced to scrub land' (Dore, 1996: 15) as social conflict within the region intensified. For Brazil's Amazonia as a whole, the main thrust of state policy 'continues to promote large capitalist enterprises. Sustainable resource use remains a low political priority' (Dore, 1996: 15).

To some extent, such seemingly 'irrational' policies may reflect economic necessity. Many Third World states remain highly dependent on the production and export of primary products. Especially vulnerable are small countries reliant on only one or two products as the basis of their livelihood: the eastern Caribbean Windward Islands on bananas, Costa Rica on beef and timber, and Ghana on cocoa, for example (Thrupp, 1990; Grossman, 1993; Kendie, 1995). However, even ostensibly powerful Third World states may be dependent in this manner. The national economy of Nigeria, for example, is 'totally dependent on the oil industry', with oil revenue comprising 90 per cent of export earnings and 80 per cent of state revenue (Rowell, 1995: 210). Since Nigerian oil production is largely controlled by Shell, the country is simultaneously dependent on the practices of this major TNC (see Chapter 5). Many Third World leaders thus have little choice but to perpetuate practices that contribute to environmental degradation in the absence of alternative sources of national income.

59

The Third World debt crisis reinforced this process. Following the onset of the crisis in 1982 (when Mexico declared a moratorium on the repayment of its debt), it soon became apparent that many Third World states had accumulated debts which they were in no position to repay in the context of a worsening economic situation (George, 1988; Adams, 1991). The result was an attempt by First World states and banks, acting mainly through the mechanism of the World Bank and the International Monetary Fund (see Chapter 4), to assist Third World states to revise 'inefficient' policies. So-called structural adjustment programmes became the norm and their adoption was typically a pre-condition for loan renewals (Reed, 1992).

The links between the Third World's debt and environmental crises are complex, but for our purposes may be considered to be twofold. First, structural adjustment programmes have usually required deep cuts in government spending to correct budgetary imbalances. The nature of such cuts has varied from country to country, but a common theme has been the reduction in the budgets of environment departments. As Dore (1996: 11) notes in a Latin American context, 'with pared budgets and reduced staffs, state agencies charged with implementing environmental regulations were unable to enforce their mandate'. That Third World states have frequently resorted to cutting the state agencies' budgets reflects the fact that these departments are usually a 'soft target' politically at a time of tough and often painful choices. Thus, the 'political constituency' of environment departments is relatively small when compared with that of other departments (e.g. forestry, transport). Further, environment departments are relatively recent creations with few of them predating 1970 in the Third World (see below). As such, they are less well established than older and more powerful departments, with the result that the latter have advantages in the bureaucratic conflicts that are inevitably associated with the allocation of budgetary cuts. Yet the budgets of environment departments are being cut at a time when the environmental crisis in the Third World is intensifying, thereby creating a dangerous and growing disjuncture between the scale of the crisis and the state's ability to respond to that crisis.

Second, structural adjustment programmes also usually require that states promote economic activities consonant with a given country's 'comparative advantage'. This emphasis on comparative advantage forms part of a much larger political and economic project associated with 'market triumphalism' (Peet and Watts, 1993) that has been linked, in turn, to the ascendancy of neo-classical economic thinking in the development of policies of the United States and other First World states since the late 1970s (Toye, 1993). An insistence on comparative advantage has also been synonymous with pressures to increase natural resource production, the mainstay of many Third World states since colonial times. This pressure for Third World states to return to 'what they do best' has not only dashed the dreams of many Third World leaders about rapid industrialisation, but has also prompted intensifying natural resource exploitation. In Latin America, for example, natural resource exports 'rose spectacularly', with fishery, forestry, mining and agricultural exports up dramatically in the 1980s (Dore, 1996). Environmental degradation has inevitably resulted in the absence of adequate environmental management safeguards. Thus, structural adjustment programmes often simultaneously reduce the ability of states to respond to environmental problems and increase the seriousness and intensity of those problems. That these programmes also contradict the message coming out of the First World that Third World states ought to promote 'sustainable development' is yet further proof of the contradictory nature of the First World's impact on the Third World (Kendie, 1995; Dore, 1996; see also Chapter 4).

The persisting imbalance between development and conservation concerns in the policies and practices of most Third World states is also linked to political factors. That national security concerns have influenced the decisions of states to encourage environmentally degrading economic activities has been noted. Such concerns, though, need not be related to external threats, but may reflect official disquiet over perceived threats to national security arising from groups within the country. The tendency to see internal threats to stability may reflect institutional 'memories' about traumatic political or economic events in the past which threatened state rule – the struggle for power following independence, for example (see above). It may also reflect more recent political or economic problems associated with, say, low-level insurgency or chronic landlessness. In Thailand, for example, the state encouraged farmers living in remote areas 'to burn down trees, so that the guerrillas could not make use of the forests' and to then grow crops in the burned areas (Rigg and Stott, 1996). This policy formed part of a larger counter-insurgency operation against the Communist Party of Thailand in the 1970s and early 1980s which sought to diffuse the perceived political and military threat of this party to the military-controlled Thai state. The policy simultaneously facilitated the assertion of state control over remote forested regions that had hitherto largely escaped central control (see, for example, Hirsch, 1990).

In a similar vein, states in many parts of the Third World have encouraged poor and marginalised farmers to migrate in their thousands, if not millions, to clear forest and settle in remote areas in part as a means to reduce social discontent in the places from which these actors have come. Thus, frontier areas in Indonesia (Kalimantan), Thailand (north-west) or Brazil (Amazonia) have served as a 'political safety-valve' to which 'surplus' landless populations have been exported, thereby obviating the need for land reform in central areas (Hall, 1989; Hurst, 1990; Weinberg, 1991; Lohmann, 1993). These 'transmigration' programmes have had the added security benefit from the state's viewpoint in that they have served to 'pacify' unruly grassroots groups living at the periphery of the country (Hurst, 1990).

The reluctance of Third World states to implement environmental conservation measures may also be related to the resistance of bureaucracies within the state itself which have benefited from the status quo. The nature of bureaucratic conflict in relation to environmental management is considered more fully below. What needs to be noted here, though, is that often the most powerful agencies within the state are precisely those agencies that have derived their institutional power from control over such environmentally damaging activities as energy generation (i.e. from coal or hydro-electric sources), intensive cash-crop production, large-scale logging or mining (Cummings, 1990; Rich, 1994). In contrast, and as noted above, environmental agencies are typically of fairly recent origin, possess little substantive power, and must confront the policies of diverse powerful agencies if they are to implement conservation measures.

Corruption among senior political leaders has often been a further political factor hindering a more 'balanced' approach to environmental management in the Third World. An important reason why many leaders fail to promote environmental stewardship at anything other than a rhetorical level is simply that it has not been in either their political or economic interest to do so. Although not all Third World leaders have benefited financially from their control over the apparatus of the state, there is nonetheless a tendency for many leaders to base political decision-making partly on calculations of personal economic gain. As Lewis (1992: 218) notes, in many parts of the Third World 'rent-seeking regimes are currently in power; some are so rapacious that they may accurately be labeled kleptocracies'. Thus, a characteristic feature of the political process in many Third World countries is the existence of a close and symbiotic relationship between state and business leaders (see also Chapter 5). An important aspect of that relationship, in turn, is the joint promotion of economic activities harmful to the environment and to the interests of those actors immediately affected by related degradation. Indeed, such a state–business 'partnership' has been the source of many of the physical changes to the environment, and related environmental conflict, that is at the heart of the politicised environment discussed in Chapter 2. Cattle ranching, logging, mining or manufacturing have all been associated at one time or another with political corruption (Hurst, 1990; Utting, 1993; Dauvergne, 1993/4). That political benefits may also accrue to leaders as a result of the judicious and highly selective awarding of contracts to exploit the environment to potential or actual political supporters must only tempt even further leaders who are keen to retain political power but who are unsure of their ability to do so without additional 'help' (King, 1993; Vitug, 1993).

State leaders thus have diverse reasons for perpetuating the political and economic status quo. It would be quite wrong nonetheless to dismiss as mere rhetoric all state initiatives formally designed to promote green development. Many Third World states have responded in recent years to calls for sustainable development by introducing 'eco-friendly' forms of capital accumulation which are leading a number of countries along what Schroeder (1995) terms 'the commodity road to environmental stabilization'. Commercial reforestation and eco-tourism are two such forms of 'green capitalism'. Chapter 2 noted that a growing number of states are confronting the problem of tropical deforestation (often linked to previous state policies) by promoting reforestation programmes centred on fast-growing commercial pine and eucalyptus. The growth of eucalyptus plantations has been especially swift as the global demand for pulp and paper has driven the industry to seek new wood supplies from the major Third World producing countries: Thailand, Indonesia, Brazil and Chile (Marchak, 1995; Lohmann, 1996). The rapid expansion of these 'pulp and paper' plantations has prompted intense social conflict locally, as well as international condemnation by ENGOs among others. For their part, participating states tout their commercial reforestation programmes as examples of how the state is fulfilling its stewardship role in society (Sargent and Bass, 1992).

Similarly, states claim green credentials for expanding the area incorporated in national parks or ecological reserves as part of their promotion of 'eco-tourism'. Tourism in general has become big business in the Third World, and eco-tourism is one of the most rapidly growing sectors within that industry (Cochrane, 1996). Third World states are capitalising on widespread tourist interest in biologically diverse tropical forests or 'exotic' wildlife through conservation initiatives that seek to protect forests and wildlife for tourist 'consumption'. Such initiatives are not new, and often find their origins in colonial times when so-called 'penitent butchers' (i.e. erstwhile big-game hunters) sponsored the creation of nature reserves and national parks (MacKenzie, 1988; Beinart and Coates, 1995). Yet it is only in recent years with the growth of a large-scale global tourist industry that tourism-related environmental conservation has really taken off. From wildlife parks in Kenya, India and South Africa to national parks in tropical forests in Costa Rica and Madagascar, much of the Third World's residual natural habitat has been enclosed in this manner (Peluso, 1993b; Utting, 1993; Ghimire, 1994; Barrett and Arcese, 1995; Kothari et al., 1995). States are able to claim that, in so far as threatened tropical forests and wildlife are protected, they are fulfilling their stewardship role.

States thus extol the environmental basis of reforestation and park development programmes. Yet political and economic factors often play the greatest role in the decision to adopt 'green' policies. To begin with, these initiatives need to be seen, in part, as a response to pressures from the First World to promote environmental conservation. Such 'eco-imperialism' is undoubtedly linked to a long-standing First World fascination with the tropical forests and their inhabitants (Putz and Holbrook, 1988; Hecht and Cockburn, 1989), and is reflected today notably in the environmental campaigns of First World-based ENGOs such as Friends of the Earth or Greenpeace (see Chapter 6). However, it is also linked to the policies and practices of First World states and multilateral institutions such as the World Bank or International Monetary Fund (see Chapter 4). Thus, with regard specifically to states, there has been a growing emphasis since the late 1980s on green conditionality in the disbursement of aid and loans from the First World to the Third World (Davies, 1992). As noted above, the general idea behind these grants or loans is that funds are released by the donor countries only when recipient countries commit themselves to pursuing policies or projects linked to green issues (Conroy and Litvinoff, 1988). To the extent that such financial transfers are now linked to environmental considerations, Third World states have a strong incentive to adopt 'green' policies (Adams, 1990).

Perhaps more importantly, most Third World states now realise that 'eco-friendly' policies can also be financially lucrative. Activities like commercial reforestation or eco-tourism may be green business (although there is growing debate on their 'greenness'), but it is the fact that these activities are also big business that is proving increasingly attractive to Third World states. Further, these growing industries have arrived at an opportune time, since traditional resource-extracting industries are in decline in many areas. These points are well illustrated in the Thai context. The Thai state is now actively promoting the spread of eucalyptus and pine plantations as part of an official campaign to reforest the country's sadly depleted forests (down to as little as 10 per cent of the total national area depending on how 'forest' is defined) (Puntasen *et al.*, 1992). It is also keenly pursuing the development of eco-tourism in the country's national parks as part of an attempt to relieve some of the pressure on Thailand's overcrowded coastal and island resorts (Ghimire, 1994). Both of these economic activities promise to deliver a substantial income to the state at a time when revenue from logging has dwindled in the measure that most of the country's old-growth forests have been felled (a situation partly offset by Thai logging in neighbouring Laos and Burma). Further, close links between businesses and the state being what they are in Thailand, these activities also augur a substantial personal income for those political leaders and state officials willing to cooperate with business. Thus, Members of Parliament from various political parties, as well as senior officials in the Royal Forest Department and the state-owned Forestry Industry Organisation, have 'a finger in the plantation or pulp business' (Lohmann, 1996: 37).

In Tanzania, meanwhile, wildlife safaris have served a comparable political and economic purpose. Long renowned as one of Africa's leading socialist experiments, Tanzania has been forced as a result of a sizeable national debt to adopt a set of pro-capitalist policies under the watchful eye of the International Monetary Fund. As 'wildlife tourism' expanded dramatically in the late 1980s and early 1990s, official revenue from this source exceeded

$100 million in 1992/3 making it a leading revenue-earner for the Tanzanian state. Concurrently, however, a series of 'land scams' associated with this lucrative economic sector have involved local politicians in illegal land transfers suggesting a process of individual enrichment by state officials not unlike that experienced in Thailand (Neumann, 1995). As these two examples illustrate, a big incentive for Third World states to 'go green' relates to the prospects for capital accumulation associated with new 'eco-friendly' industries.

Finally, states may also be promoting environmental conservation for reasons related to internal security or social control. Conservation initiatives may thus be a means for states to assert their authority over peoples and environments hitherto subject to weak control, thereby strengthening the position of the state in relation to other actors (even if grassroots 'participation' is ostensibly part of the process: see Ribot, 1995). The creation of a national park or an eucalyptus plantation, for instance, almost invariably involves systematic state intervention in the lives of the people living in the designated area (Neumann, 1992; Ghimire, 1994). In the case of the creation of a park, there is not only the delimitation of the borders of the new administrative entity, but also the appointment of a whole 'army' of park rangers and guards to ensure that excluded actors do not interfere with park management. These officials also keep potential 'troublemakers' in the area under close surveillance in a manner akin to a military operation. That this process is not simply a 'technical' matter can be seen by the fervent opposition put up by local farmers or shifting cultivators to the creation of parks or ecological reserves in countries as diverse as Costa Rica, Tanzania, Thailand and India (Utting, 1993; Peluso, 1993b; Kothari et al., 1995; Lohmann, 1996). In the Indian context, official efforts to protect tigers, elephants and other prized mammals have usually involved the displacement of local people whose 'way of life is viewed as inimical to wildlife conservation' (Kothari et al., 1995: 191). There is thus a strong coercive element to the development of 'green' projects in India and elsewhere in the Third World as states use force, where necessary, to protect valued wildlife and trees, as well as to crush opposition from grassroots actors.

Environmental conservation is therefore rarely seen by states as an end in itself, but rather as a means to various political and economic ends. Yet, whether relating to activities that lead to environmental degradation or conservation, the state is an actor that rarely speaks with one voice, but rather represents an amalgam of institutional interests. Under such circumstances, the tension between the state's role as developer and steward of the environment often plays itself out in terms of conflict between rival agencies within a state, or between different states interacting at the international level.

INSTITUTIONAL CONFLICT

As the 'state' encompasses diverse agencies and interests, and further operates in a world of states, the state's role as an environmental manager needs to be understood as an outcome of intra-state, as well as inter-state, conflict. The importance of state attributes and decision-making patterns to an understanding of broader political-ecological issues is thereby highlighted.

An important starting point in attempting to understand the nature and significance of intra-state conflict is the development during the colonial era of the functionally differentiated state, and the associated institutionalisation of conflict over environmental matters within the state. A series of departments were created by the colonial powers during the nineteenth and early twentieth centuries that specialised in a given aspect of environmental management: forestry, agriculture, fisheries or mining. These functionally defined departments were not created all at once, but rather incrementally as funds became available, and as the need arose (as one would expect, key departments such as agriculture and forestry were typically established first). In some cases, hybrid departments were created in which two or more functionally defined services were combined within the same department – for example, Agriculture and Forestry. However, notwithstanding the vicissitudes of organisational expression, the overall aim was to conceptualise and manage natural resources in a functionally defined manner so as to maximise their commercial return (Bryant and Parnwell, 1996).

To appreciate the full significance of this administrative change, it is necessary only to note the structure of the state in precolonial times. Precolonial states varied from place to place, but there usually existed a general structure in which the ruler claimed authority over all people and resources within a territory, but in practice had to share such authority with other actors. In Burma, for example, the king shared political power with regional and local notables who were often very powerful actors in their own right (Leiberman, 1984). Resource management under this system was a highly personalised affair as individual notables derived their wealth and power from the right to tax local resource use, which was held, in theory at least, at the sufferance of the ruler. Thus, it was the individual's relationship to the ruler, and not the possession of specialised resource knowledge per se, that determined who controlled resource management in precolonial times. The functionally defined state was designed in part to get away from what colonial rulers considered to be a precolonial system of environmental management based on nepotism rather than rationality.

This ostensibly innocuous administrative change transformed the way in which states managed the environment in the Third World. The new functional departments enhanced 'efficient' resource extraction, as measured in a series of quantitative indices related notably to resource production and revenue levels (Bryant, 1996a). However, the change also magnified the state's conflicting roles as developer and steward of the environment, and placed this conflict at the heart of the state's environmental policies and practices. In a sense, the creation of functional departments was partly an attempt by colonial rulers to reconcile maximum resource extraction (and revenue) with the maintenance of long-term supplies in the case of potentially renewable resources (notably timber). Early nineteenth-century colonial policies in South and South-East Asia based on *laissez-faire* principles had only led to rapid resource depletion, as well as to fears of resource scarcity and irreversible environmental decline. By the mid-nineteenth century, therefore, many colonial states had recognised the need to strike a balance of some sort between resource exploitation and conservation (Grove, 1990; Peluso, 1992; Bryant, 1994b).

The new functional departments proved singularly incapable of achieving such a balance. Indeed, the reorganisation of the colonial state along functional lines set in motion a process of institutional development in which state environmental management became highly fragmented and prone to conflict. Thus, as functional departments were established, professional training became a prerequisite for entry to service. This step enabled departments rapidly to acquire specialist knowledge about the resource in question and, crucially, about how best to maximise production levels. Yet it also encouraged parochialism in so far as officials now remained in a given service throughout their careers and thus had little or no experience of the other services or of how their own work fitted into a broader picture. As Furnivall (1956: 77) remarked, 'none of these [specialist] officials saw life whole and, by reason of frequent transfers, none of them saw it steadily'.

This situation frequently led to bureaucratic conflict as departments fought with each other (but also with grassroots actors, see Chapter 7) in the course of pursuing their duties. A basic disjuncture thus developed between the ways in which the colonial state organised its administrative services to manage environmental resources on the one hand, and the actual condition of the resource base itself on the other hand. In effect, the 'political/administrative world' did not coincide with the 'real resource world' that the former sought to administer (Bryant and Parnwell, 1996). Thus, the real resource world did not conform neatly to official resource categories (i.e. 'forests', 'agriculture'), but rather overlapped categories in complicated ways, thereby virtually guaranteeing bureaucratic conflict in a state organised along functional lines.

The relationship between agricultural and forestry officials in late nineteenth- and early twentieth-century British Burma illustrates this point. Here, agriculture officials presided over one of the most rapid land clearances anywhere in the Third World in the late nineteenth century as more than three million hectares of forest were cleared for permanent agriculture by farmers in southern Burma alone (Adas, 1983). These officials measured progress in terms of hectares of forest cleared and planted with cash crops (mainly rice), and promotion was linked to this process. In contrast, forestry officials sought to protect commercial teak forests from encroachment by an advancing 'tide' of farmers. These officials defined progress in terms of hectares of commercial forest protected for long-term timber exploitation and here too career prospects hung in the balance. Conflict was all but inevitable under the circumstances; agriculture and forestry officials disputed which forests to protect or abandon, and the rules that were to govern the access of farmers and shifting cultivators to those forests which were to be protected (Bryant, 1996c).

Such intra-state conflict has persisted in the postcolonial era as Third World leaders opted almost without exception to retain the functionally defined state after independence (states in Latin America had already adopted this organisational form during the late nineteenth and early twentieth centuries). Indeed, a dramatic expansion in the developmental role of the Third World state following the Second World War resulted in further differentiation along functional lines (Hirsch, 1990; Pathak, 1994). The new departments or agencies created at this time enhanced the state's already substantial capacity to manage and degrade the environment, but also led to increased intra-state tensions over the state's development and stewardship roles.

Resource departments keen to maximise their power within the state hierarchy continue to squabble over resource exploitation issues today. As in colonial times, the issue with these departments is not whether or not to exploit the environment. Rather, it revolves around the relative priority to be attached to the different activities that comprise such exploitation when different resource uses conflict with one another. The conflict between agricultural and forestry departments has persisted wherever residual forest is a likely candidate for conversion to permanent cash-crop agriculture (Rush, 1991; Pathak, 1994). Other intra-state conflicts have developed in postcolonial times as the pressures to maximise environmental resource use intensify. In Burma, for instance, conflict has developed between mining and forestry interests within the ruling State Law and Order Restoration Council (the military junta presently in charge of the country) over whether to permit environmentally damaging mining in valued forests (Bryant, 1977a). In neighbouring Thailand the Royal Forest Department and the Tourism Authority of Thailand compete for control of that country's national parks (Handley, 1994). In so far as conservation becomes an issue in these conflicts at all, it is over whether a specific commercially valuable resource is to be protected for long-term use, rather than about the conservation of the environment per se.

Yet as this chapter has already noted, most Third World states have responded to the intensifying environmental crisis within their borders by creating 'environment' departments and agencies specifically dedicated to promoting environmental conservation. In this manner, growing concern over the environment has been reflected in the institutional structure of the functionally defined state itself. This step has added a new dimension to intra-state conflict, pitting a department or agency intent on managing and protecting the environment as an integrated entity against traditional resource departments keen, for the most part, to maintain the existing compartmentalised approach. Not surprisingly, environment departments and agencies have rarely emerged victorious from this conflict, and have been habitually relegated to a secondary role within the state's decision-making hierarchy. This outcome partly reflects the fact that resource departments have usually been around much longer than their environment counterparts, and hence can draw on a much more extensive network of political contacts than their colleagues in environment departments or agencies. Such contacts are usually crucial to the outcome of inter-bureaucratic conflicts.

The case of the Indonesian Environmental Impact Management Agency (*Badan Pengendalian Dampak Lingkungan Hidup* or BAPEDAL) may be briefly illustrated here. Established by presidential decree in June 1990, BAPEDAL is a national environmental regulatory body with a brief to assist the president in promoting 'better environmental management' practices in the country. However, the performance of the agency so far has been disappointing, with the existing bureaucracy a major obstacle to BAPEDAL's development: 'The Indonesian bureaucracy is large and slow in its procedures, and is characterised by a marked tendency to favour compartmentalisation and fragmentation of responsibilities rather than coordination. The impact of these factors on BAPEDAL's development is evident' (MacAndrews, 1994: 99). The limited impact of the agency also reflects the fact that BAPEDAL has advisory powers only, 'with its effectiveness therefore depending on its access to and influence with senior government personnel' (Eldridge, 1995: 139).

The weakness of an environmental agency such as Indonesia's BAPEDAL also reflects the essential incompatibility of this type of agency with the internal logic of the functionally defined state. In effect, it attempts to bridge the gap between the political/administrative world and the real resource world that developed in colonial times with the advent of this type of state. Yet, without external political support (and this has often been lacking – rhetoric notwithstanding), environment departments and agencies are inevitably doomed to fail in this task precisely because success would seemingly entail the end of the functionally defined state itself – a highly unlikely prospect given the power of the latter's supporters both inside and outside the state.

The discussion so far has explored the intra-state dynamics that influence the Third World state's role as developer and protector of the environment. However, that role has also been conditioned by the relationship of individual states to each other in an international system characterised by widespread conflict over environmental issues. We suggested earlier that a 'tragedy of the commons' scenario might exist at the international level as a result of states acting individually in the pursuit of their own national interests. The relationship between state sovereignty, state policies and international environmental change is complex (Porter and Brown, 1991; Lipschutz and Conca, 1993; Vogler and Imber, 1996). However, for our purposes two aspects to that relationship can be noted briefly here.

First, international environmental degradation is partly an outcome of the fact that states create environmental policies and practices within the territories under their control, but those policies and practices may generate an environmental impact that goes well beyond the territories in question. This disjuncture between political responsibility and possible environmental effect applies to all states in the world, but what is of interest here is the link between rapid industrialisation in selected Third World countries, including the policies of the states that encourage such development, and the growth of acute trans-boundary pollution problems at the regional and global scales. Thus, the creation of 'pollution havens' in selected Third World countries has been associated with the growth of international air and water pollution problems which may manifest themselves most vividly in selected regions (e.g. South-East Asia, southern Brazil), but which also ultimately contribute to global environmental problems (McDowell, 1989; Hardoy et al., 1992).

As the role of the Third World in global environmental pollution has grown, however, calls in the First World for Third World states to take action to curb greenhouse gas emissions have increased (World Resources Institute, 1990). Yet whether the Third World is ultimately to blame for such pollution, as well as whether the Third World can or should be expected to pay for costly environmental clean-ups, is the source of growing mutual recrimination and conflict between the Third World and the First World (Agarwal and Narain, 1991; Porter and Brown, 1991). From the United Nations Conference on the Human Environment held in Stockholm in 1972 to the United Nations Conference on Environment and Development held in Rio de Janeiro in 1992, such conflict has been a recurring feature, and has been a factor contributing to the failure of these and other international conferences to effect significant change in the policies and practices of the world's states (Middleton et al., 1993; McCormick, 1995; see also Chapter 4). The unwillingness of Third World states to modify policies that promote industrial development at the expense of the environment has long been matched by the opposition of First World states to any strategy that would provide funds to poor Third World states to enable them to introduce costly environmental clean-up programmes – although the First World's recent agreement to establish a fund for Third World countries relating to the phasing-out of chlorofluorocarbons (which contribute to ozone depletion) suggests that the log-jam may finally be breaking (see below).

Second, international environmental degradation is also associated with the policies and practices of states acting outside their own national boundaries in international 'commons' (Vogler, 1995). Here again, conflict between states has been a frequent occurrence, in this case over access to environmental resources such as minerals, whales or fish. Various international agreements or 'regimes' have been devised to end or prevent inter-state conflict and resource over-use, notably with reference to whaling and polar environmental management (Young, 1989; Hurrell and Kingsbury, 1992). Such agreements or regimes, however, are often quite fragile: international efforts to ban whaling on the high seas to prevent the extinction of endangered species has long been undermined by the actions of pro-whaling states in both the First and Third Worlds, for example (Porter and Brown, 1991; Stoett, 1993). Further, states have not been able to come to an agreement at all on other international resource issues such as the protection of dwindling global fish stocks. As these stocks decline, large-scale fishing operations (often sponsored by states keen to maintain domestic fish supplies) have spread throughout the Third World's fisheries, prompting conflict between states as well as between transnational and local fishers (Fairlie, 1995). The case of inter-state conflict over international resource use thus highlights the role of states in contributing to international environmental problems on the one hand, and the limits to coordinated state action to attempt to resolve those problems on the other hand.

It would nonetheless be wrong to suggest that the prognosis for inter-state cooperation is all bleak (Caldwell, 1990; McCormick, 1995). In certain cases, agreements have been reached – the Montreal Protocol on Substances that Deplete the Ozone Layer of November 1987 is a case in point. Under this agreement, states have agreed to phase out the production of ozone-depleting substances over a specified time-scale (Benedick, 1991). A major source of conflict in the negotiations surrounding this agreement, and its implementation, was the question of compensation to industrialising Third World countries for the proposed phase-out (Porter and Brown, 1991; Litfin, 1994). However, the pressure of Third World states led by India, China and Brazil prompted First World states (led by the United States) to reverse their position in 1990 and agree to a fund for Third World countries. In effect, this policy reversal reflected a First World recognition that the Third World's cooperation was 'essential to the long-term success of the Montreal Protocol', and that such cooperation would not be forthcoming in the absence of such a fund (Miller, 1995: 80).

The general unwillingness of states to surrender sovereignty, as noted earlier, nonetheless remains a perennial stumbling block to conflict resolution over environmental management issues at the global scale (Johnston, 1992). Thus, the recent upsurge in international agreements as a result of the Rio Earth Summit belies the fact that 'there is a marked preference for non-binding targets/guidelines which states are free to implement at whatever pace they see fit rather than the acceptance of firm and unambiguous obligations' (Hurrell, 1994: 152). In the absence of an effective and equitable global environmental management regime, therefore, most Third World states (as with their First World counterparts) persist with policies that favour economic development over environmental conservation goals, even if those policies contribute to global environmental problems (Centre for Science and Environment, 1992).

The preceding discussion of institutional conflict has shown how intra-state and inter-state conflict can affect the Third World state's environmental management role. It has suggested that, just as the functionally defined state has proved a major obstacle to environmental conservation efforts within a country, so too the territorial definition of states has so far prevented the emergence of effective global environmental cooperation designed to tackle environmental problems at that scale.

THE DECLINE OF STATES?

This chapter has painted a fairly bleak picture of the role of the state as an environmental manager in the Third World. It has argued that while states have grown in power over the centuries this trend has served only to enhance environmental degradation because states have used their political power to promote economic development over environmental conservation. While many (but not all) states have come to play the pivotal role in human affairs predicted by Hobbes in 1651, that role has been one largely associated with destroying, rather than conserving, the environment. Thus, many of the political-ecological problems referred to in this book can be traced directly or indirectly to the policies and practices of states.

What remains to be considered in this chapter, though, is whether Third World states are likely to continue to play a pivotal role in political-ecological issues in the future. In a world characterised by increasingly powerful non-state environmental actors, it is far from evident that Third World states will continue to hold the privileged position that many of them have enjoyed over the past fifty years or so. To be sure, not all states in the Third World have been as powerful in practice as they have claimed to be in theory (Migdal, 1988). As this chapter has suggested, in many cases the contemporary Third World state has nonetheless played a central role in mediating human–environment interaction. However, that privileged position may now be under threat as a result of the growing power of other actors linked to the combined forces of 'globalisation' and 'localisation'.

On the one hand, the development of a globalised capitalist system since the Second World War has seen the rise of the transnational corporation (TNC) as a major economic actor on the global stage (Gill and Law, 1988). The scope and nature of the economic power of this actor is considered in

Chapter 5. Here, the concern is with the ability of this actor to weaken the power of states in the Third World to manage affairs within their own territories.

A key feature of 'globalisation' has been the integration of most parts of the world into a single economy dominated by TNCs. Now, it is in the very nature of a TNC to have operations in diverse countries with the location of a firm's operations determined by such factors as the availability of environmental resources, cheap labour and local markets (Korten, 1995). As noted, many Third World states have sought to industrialise rapidly, and a common way in which they have sought to do so has been to attract TNCs to their country through the creation of 'favourable' business conditions that have included, among other things, weak environmental regulations (Leonard, 1988; Hardoy *et al.*, 1992). To the extent that states thereby abdicate their stewardship role in the pursuit of TNC investment, the ability of TNCs to translate economic might into political-ecological power is demonstrated.

Globalisation has resulted in a more general erosion of the ability of states in the Third World to dictate the pace and nature of economic development, and by extension, environmental conservation. As these states become dependent on trade and investment linkages with the outside world, they often lose the ability to control the development process. Day-to-day decisions which have an important social and environmental effect within a country are accordingly as likely to emanate from the headquarters of TNCs in Tokyo, London and New York (and now also increasingly Taipei, Seoul and Hong Kong) as they are to be made by political leaders in the national capital.

Yet the power of TNCs vis-à-vis Third World states should not be exaggerated. In the case of the more powerful states such as Brazil, Indonesia or China, state leaders retain considerable power to regulate the development process, and thereby those TNC operations that affect the country. Indeed, those leaders are often important economic players in the national economy themselves – in addition to their political functions (see above). This concentration of political and economic power provides a counterpoint to the ability of TNCs to dominate the political process in many Third World states. However, the fact that even relatively weak Third World states possess ultimate formal authority over national affairs – and that TNCs must obtain the permission of the state to establish and maintain operations in a given country – ensures that the relationship between states and TNCs is rarely completely one-sided (Pearson, 1987).

On the other hand, intensifying social and environmental crises in many parts of the Third World have prompted the development of an increasingly assertive 'grassroots' movement which, acting in conjunction with First and Third World ENGOs, demands the devolution of powers from the state to the local community level (Ekins, 1992). Chapters 6 and 7 explore this process in greater depth, while it is the impact of 'localisation' on the capacity of the state to act which is of concern here.

If TNCs constitute a potentially powerful challenge to Third World states as a result of their immense economic power and ability to influence the course of economic development, then grassroots organisations (in association with ENGOs) represent an equally great threat to the authority of states to manage people and environments within a territory. However, while TNCs have developed as part of the globalisation of economic activity, grassroots groups typically develop as a local response to the perceived shortcomings of the state – notably, its failure to promote social equity and environmental conservation. The challenge of these actors to the state thus resides in an ability to question the legitimacy of the state as the main promoter of society's social and environmental interests.

The power of grassroots groups resides primarily in their ability to hinder development projects supported by the state, but which are the source of local popular opposition. As Chapter 7 shows, this ability is not to be gainsaid, for it has led to the termination of a growing list of projects in the Third World. Yet the power of grassroots actors should not be over-rated either. Thus, although their power is often linked to detailed local social and environmental knowledge, and a tenacious pursuit of local interests, those interests are far from homogeneous, with the result that serious differences may develop over their definition (Guha, 1989). Further, their ability to challenge broader power structures is weakened as a result of the 'Achilles heel of localization' (Esteva and Prakash, 1992). To some extent, ENGOs may help to overcome this weakness by facilitating the exchange of ideas and coordinating common action on social and environmental issues. Often able to muster international financial support and media coverage, these ENGOs simultaneously publicise the 'legitimate' claims of grassroots actors and the perceived deficiencies of state policies and practices (see Chapter 6).

A tactical alliance between grassroots groups and ENGOs is thus a potentially potent means of challenging the environmental policies and practices of the Third World state. Yet, if these actors are able to highlight the perceived deficiencies of state actions, they are scarcely in a position to replace completely the state – either as a promoter of economic development or as a protector of the environment (Hurrell, 1994). Further, the devolution of power to grassroots actors would be no guarantee that such localised power would be used to promote sustainable environmental practices (Bebbington et al., 1993). Only the state, moreover, is in a position to mediate between the competing pressures of 'globalisation' and 'localisation' as they come to bear in a given locality.

Thus, it would appear premature to write the obituary of the state as a key actor in environmental management and conflict in the Third World. In the words of Hobsbawm (1996: 276), the state 'remains indispensable' today to the functioning of modern societies. At the junction of the 'global' and the 'local', the state is the only actor today in a position to address with authority political and ecological problems at a variety of scales (see Chapter 2). That the state has often failed to do so – indeed, has been an important agent contributing to such problems – in practice reflects, as this chapter has shown, a long history of the state's favouring economic development over environmental conservation. Further, the very manner in which the state has come to be functionally organised may be a major obstacle to reforming the state so as to improve its ability to protect the environment. The next chapter considers whether multilateral institutions offer a way around this problem in so far as these institutions are able to deploy sizeable technical and financial resources to assist states in the resolution of the Third World's environmental problems.

4

MULTILATERAL INSTITUTIONS

A key feature in the development of the global capitalist economy in the twentieth century has been the creation of a network of multilateral institutions whose primary aim has been to promote social and economic development through the provision of technical and financial assistance. A group of international financial institutions led by the World Bank and the International Monetary Fund (IMF) was created to assist financially and regulate the development programmes of less developed states, while a different set of technical institutions such as the Food and Agriculture Organisation (FAO) and the Industrial Development Organisation (UNIDO) was founded under the auspices of the United Nations to serve a technical support function. Not surprisingly, given these remits, the multilateral institutions have focused much of their energies on the Third World, and their collective social and environmental impact in this part of the world has been immense.

This chapter examines the contribution of multilateral institutions to Third World environmental change and conflict in order to gauge the role of these institutions in the Third World's politicised environment. The rise of multilateralism in the aftermath of the Second World War can be seen from one perspective to be a political response to that war – viz. the creation of a 'United Nations' system as a potential bulwark against the prospect of a 'third world war'. However, what is of particular interest here is the way in which multilateral institutions developed into leading proponents and sponsors of 'development' within the global capitalist system – with all the attendant social and environmental consequences. Thus, technical organisations led by the FAO have sought to provide 'technical' input into the decision-making process of Third World states on diverse natural-resource development and manufacturing issues. In contrast, the World Bank and other international financial institutions have proffered financial assistance and 'technical' advice on both specific development projects, and, through structural adjustment programmes, on macro-economic questions as well. Yet, as political ecologists point out, the activities of these multilateral institutions have been strongly imbued with political and ecological meaning. Far from being the neutral 'technical' assistance that proponents claim, the activities of multilateral institutions have often been at the centre of conflict over the Third World's politicised environment.

THE RISE OF MULTILATERALISM

The emergence of multilateral institutions dedicated largely to the provision of technical and financial assistance to the 'developing' Third World needs to be linked to the broader process of growing multilateralism in world affairs in the late twentieth century. A striking feature about contemporary international affairs is the increased frequency and political prominence of world commissions and conferences, an outcome of which has been the apparent strengthening of existing multilateral institutions, and even the creation of new institutions (e.g. United Nations Commission on Sustainable Development). Political ecologists have emphasised two things in this regard. First, they question whether these meetings can be anything other than 'talking shops' in a world characterised by the highly unequal power relations that multilateral institutions themselves embody. Second, they argue that, in so far as multilateral institutions have tended to promote development within the existing political and economic order, these institutions are supportive, in practice, of precisely those social and economic inequalities that they are committed to eradicating in theory. Further, the practices of these institutions themselves have been a major factor behind environmental degradation in the Third World. In short, political ecologists suggest that the major multilateral financial and technical institutions are frequently part of the problem, and not the solution, in terms of the Third World's environmental crisis.

The origin of contemporary multilateralism, in the context of the need to find a mechanism by which to regulate increasingly destructive conflict between states, has been a long-standing theme in international relations literature, with attention in recent years being devoted to the economic significance of multilateralism in a global capitalist system (Keohane, 1986; Gilpin, 1987; Murphy and Tooze, 1991; Walker, 1993). What is of interest here is the way in which multilateralism has been linked to the promotion of economic 'development' in the 'Third World' and how, in so doing, it has been associated with economic activities that have frequently produced widespread social hardship and environmental degradation (Escobar, 1988; Ferguson, 1990; Peet and Watts, 1993).

The ways in which specific multilateral institutions have been linked to Third World environmental change and conflict is explored below in relation to a selective analysis of the activities of key institutions. It is useful here, however, to explore briefly the general impact of these institutions on the evolution of the Third World's politicised environment. In this regard, four themes stand out. First, the major multilateral institutions have operated on the basic principle that the capitalist system is the appropriate way in which to organise economic activities worldwide. Indeed, the overall effect of their activities can be broadly described as having contributed to the introduction of capitalist practices around the world. The role of the international financial institutions in this process is most evident since these institutions are directly involved with questions of economic organisation and activity. The way in which the latter promote capital accumulation has changed – broadly speaking from a Keynesian state-interventionist stance between the late 1940s and the late 1970s to a neo-classical pro-market stance since the early 1980s – but this shift should not obscure the central goal of encouraging such accumulation (Krasner, 1985; Toye, 1993). Less evident, but no less significant, is the way in which technical institutions like UNIDO and the FAO facilitate capitalist accumulation through the provision of the requisite technical support. Thus, the FAO has been a keen promoter of intensive forestry and agricultural techniques designed to maximise food and fibre production for the global market (Westoby, 1987). This goal may have been designed to meet the rapidly growing needs of the world's populations, but the means by which those needs were to be met – the global capitalist market – were never in doubt. As Chapter 5 illustrates, however, that market has been associated with intensified natural-resource extraction and manufacturing activities that lie at the root of the Third World's environmental crisis.

Second, multilateral institutions have generally been deeply supportive of the political position of Third World states in society. That support has certainly waxed and waned somewhat according to the underlying economic philosophy guiding the multilateral institutions (and their First World supporters). Thus, when Keynesian interventionism was the order of the day support was especially strong but, since the triumph of neo-classical pro-market thinking in the 1980s, multilateral institutions have moderated their support for Third World states (Toye, 1993). Yet the latter process (discussed further below) should not obscure the basic support of multilateral institutions for states, and for the central role of states in the development process. After all, these institutions are the creation of states, and are still financed by states, and it would be unusual, to say the least, if they were to go against the fundamental political interests of those actors. For example, multilateral institutions take as read the fact that the authoritative form of political organisation today is the nation-state, and that their main 'customers' in the Third World are states. Indeed, the functional definition of the state (see Chapter 3) is reflected in a comparable functional approach by multilateral institutions, the better to coordinate the activities of these institutions. Thus, the state sovereignty that we saw in the last chapter to have been an important obstacle to global environmental cooperation has been supported generally by the multilateral institutions. As noted below, the activities of these institutions are often a powerful force in aid of state sovereignty and power.

Third, multilateral institutions are increasingly supportive of businesses (especially transnational corporations: TNCs) in the Third World. One aspect of the triumph of the pro-market view in recent years has been the quest by multilateral institutions, led by the World Bank and the IMF, to encourage less state intervention in the economy in favour of local and transnational businesses (see Chapter 5). While these institutions remain highly supportive of the political position of states, they have been increasingly scathing about the economic interventionist strategies that many Third World states have pursued over the years (it must be said, often with the active support of many of those same multilateral institutions). Not all multilateral institutions share with equal fervour the pro-market approach: that approach is especially influential in the international financial institutions, but is perhaps less favoured in the technical institutions. However, the shift towards a pro-market stance is under way throughout the multilateral sector and, in the process, the institutions in this sector are under growing pressure to link their activities explicitly to the promotion of business opportunities and interests (Watts, 1994; Korten, 1995).

Finally, the major multilateral institutions have pursued policies that have contributed to the marginalisation of millions of grassroots actors in the Third World in so far as the vision of capitalist development that these institutions espouse has been predicated on the systematic and widespread enclosure of land and other environmental resources used by these local actors. Chapter 3 described how Third World states have pursued development in such a manner as to earn themselves a general reputation as prime 'destroyers' of the environment. Yet, the ability of these states to effect such destruction on a large scale is related, in part, to the financial and technical resources placed at their disposal by multilateral technical and financial institutions (George and Sabelli, 1994; Rich, 1994). The subsequent imposition of structural adjustment programmes has only exacerbated this process (see below). As this chapter makes clear, multilateral institutions have rarely, if ever, been on the side of poor and marginal grassroots actors – rhetoric notwithstanding (*Ecologist*, 1993).

The remainder of this chapter examines in more detail the implications of these general remarks in relation to the practices of the major multilateral institutions. First, however, it is important to say a few words about the world commissions and summits that occur today with growing frequency, and which broadly reinforce the position of multilateral institutions. There is a growing literature on the nature and implications of environment and development world 'summitry' as a form of global management (e.g. Adams, 1990; Caldwell, 1990; Middleton *et al.*, 1993; Sachs, 1993; Chatterjee and Finger, 1994; McCormick, 1995). Here, a brief discussion linked to the political-ecological issues of this book is appropriate.

The growth of world summitry is related to growing worldwide public and official concern about a whole host of social and environmental problems that are manifesting themselves increasingly at the global scale. Yet the specific form and content of world summits reflects clearly the unequal power relations discussed throughout this book. As political ecologists are wont to point out, environment and development world summits – viz. the United Nations Conference on the Human Environment (held at Stockholm in 1972) and the United Nations Conference on Environment and Development (held in Rio de Janeiro in 1992) – are long on First World environmental concerns and short on Third World development problems (Hecht and Cockburn, 1992; Middleton *et al.*, 1993; Sachs, 1993; Chatterjee and Finger, 1994). The point is not that issues of Third World poverty and environmental degradation were absent from these summits – quite the reverse. Rather, it is that First World states (led by the United States) use their considerable political and financial leverage over most Third World countries to ensure that the 'global agenda' remain firmly a 'First World agenda'. Thus, those environmental topics, such as oceanic pollution, global warming or biodiversity loss, of most concern to the First World, figure prominently in summit deliberations (and any associated agreements) (Miller, 1995). In contrast, those development issues, such as First World resource over-consumption, global trade imbalances, and the causes and remedies of Third World poverty, of principal interest to many Third World states, are marginalised in the conference proceedings. There are certainly variations on this theme depending on the topic in question. Further, state interests do not always conform neatly to a First/Third World dichotomy (Williams, 1993). However, the broad thrust of our argument here is true enough.

Unequal power relations among actors is also reflected in the vexatious issue of attributing causation (and hence blame) in the development of global environmental problems at world summits. First World states tend to favour the view that many of those problems are linked to problems emanating from the Third World – above all, to the 'population problem' (Harrison, 1993; see also Chapter 1). Their view is that 'runaway' population growth in most of the Third World contributes simultaneously to social poverty and environmental degradation. However, as the perceived link between tropical deforestation and global warming illustrates, the Third World's problems are now the world's problems, and this situation has been used to justify growing First World interference (via bilateral and multilateral mechanisms) in Third World social and environmental matters (Sachs, 1993; Miller, 1995; and see Chapter 3). In contrast, many Third World states emphasise the contribution of First World states to the Third World's social and environmental problems, but summit proceedings have never even remotely suggested that the Third World be permitted to intervene in First World matters (Chatterjee and Finger, 1994).

Finally, unequal power relations at world summits are also highly visible in terms of the proposed solutions to global environmental problems, and which actors are to be empowered as a result of those solutions (see also Chapter 2). The summits have tended to enhance the power of multilateral institutions under the control of the First World, such as the FAO or the World Bank (see below). The summits have also served the wider purpose of affirming the existing political and economic order of things. Thus, they have confirmed the central role of states as the 'building blocks' of global environmental management initiatives (despite the evident problems with that role, see Chapter 3). Further, and in the 'pro-market' 1990s, the Rio Conference revealed the ability of businesses, especially First World-based TNCs, to translate their immense economic power into political rights associated with their new-found status as global promoters of sustainable development (see Chapter 5). In contrast, the poor and marginal grassroots actors, who are typically the worst affected by the various social and environmental problems considered at world summits, remain objects of discussion rather than subjects whose interests are actively promoted through grassroots-based solutions (see Chapters 6 and 7). Thus, what has been termed the emerging 'global environmental governance system' (CSE, 1992: 262) is not set up in such a way as to promote the priorities of grassroots actors, but is rather designed to promote a 'sustainable development' defined and directed by political and economic elites in both the First and Third Worlds. As espoused through this emerging global environmental governance system, the word 'global' does not denote a universal human interest, but rather signifies 'a particular local and parochial interest which has been globalized through the scope of its reach' (Shiva, 1993: 150).

Thus, the rise of multilateralism as manifested through the creation of multilateral institutions and world summits would appear not to denote a quest for an equitable collective solution to the diverse environmental problems facing humankind today. Rather, it would appear to be a phenomenon closely allied to the worldwide promotion of 'parochial' interests linked to states in both the First and Third Worlds and, increasingly in the 1990s, also to global business interests. It remains to specify in more detail the ways in which key multilateral institutions, under the strong influence of powerful First World states, play a role in Third World environmental change and conflict, a subject to which we now turn.

UNITED NATIONS TECHNICAL ORGANISATIONS

At first glance, the United Nations technical organisations established in the wake of the Second World War appeared the very embodiment of an ability to approach complex social and environmental problems from a technical and non-political viewpoint. Thus, the FAO concerned itself with the promotion of 'rational' agricultural, forestry and fisheries policies in member states, the International Labour Organisation (ILO) addressed itself to labour training and development policies, while UNIDO attended to the promotion of industrialisation, notably in the Third World – to name but three examples. In addition, special United Nations 'Economic Commissions' were established in Latin America, Africa and Asia to tailor technical advice and training to specific regional needs. However, as the following examination of the role of the FAO in Third World environmental change and conflict shows, these technical organisations have hardly been non-political in their policies.

The FAO was established in 1945 primarily to 'develop world agriculture so to enable the World to feed itself' (Sesmou, 1991: 47). This ambitious aim was to be met through a series of measures that would 'modernise' the agricultural sector, promote the export of cash crops, and enhance productivity and 'efficiency' in this vital sector generally. However, the FAO's role in this process was largely indirect, and revolved around the provision of technical advice and support to states and development agencies responsible for planning agricultural development. Thus, 'as an institution, FAO derives its power and influence not from the funds it controls, which by the standards of the international development community are trifling, but rather from the role it plays in providing technical advice and assistance' (Marshall, 1991: 66). That role has been far from inconsequential in that FAO advice and training on how best to promote 'efficiency' in the agriculture, forestry and fisheries sectors has been an important influence on patterns of environmental use and conflict in the Third World.

It was almost inevitable that the FAO would focus much of its attention on the Third World. Not only was this part of the world home to the vast majority of humankind, and the location where population numbers were growing the most rapidly, but it was also the region considered to be most desperately in need of 'modernisation' (Peet, 1991; Hettne, 1995). The strong links of the FAO to colonial rule (notably in Africa where the FAO worked alongside colonial counterparts in some cases into the 1960s) only strengthened its institutional view that the natural resource management strategies of grassroots actors throughout the Third World were 'backward' and 'inefficient' (Richards, 1985). As in colonial times, 'progress' for the FAO was defined in terms of increased production for the market, and modernisation was thus the process whereby Western science and technology were applied to the task of enhancing yields (Jewitt, 1995; Bryant, 1996a). This quest to apply Western knowledge in aid of 'development' mirrored that of the World Bank and other international financial institutions (IFIs) discussed below, but influence for the FAO was derived almost entirely from the transmission of ideas of 'good practice', rather than through the transferral of funds. Yet what is striking about the FAO since 1945 is how little that institution's core ideas (noted above) have changed. Thus, in an important policy document first published in 1979 (*World Agriculture: Toward 2000*), the FAO remained adamant that agriculture in the Third

82

World needed to be 'modernised' (Sesmou, 1991). Indeed, in the late 1980s, the FAO was insisting that, 'without modern agricultural inputs, Africa will be seriously constrained by being locked into traditional production technologies' (cited in Sesmou, 1991: 48). In the fisheries sector, meanwhile, the emphasis was on increasing the capacity of fishers to increase catches through more technologically 'sophisticated' fishing vessels (Fairlie, 1995). The concern in the forestry sector was largely to promote the spread of commercial forest plantations so as to meet rapidly expanding world demand for timber as well as pulp and paper (Marchak, 1995).

In each of these cases, a notable outcome of FAO-supported intensive resource use has been deepening environmental degradation and the social marginalisation of poor grassroots actors. As Chapter 7 documents, this process has prompted fierce environmental conflicts in most parts of the Third World. Concurrently, however, the local environmental knowledge possessed by these actors was dismissed by FAO officials and other 'scientific experts' as inferior to 'modern' Western knowledge systems. Yet, as political ecologists have pointed out, the latter are only one way among many others of understanding the environment, and the place of humanity in relation to that environment (Richards, 1985; Redclift, 1992; Banuri and Marglin, 1993). The conflict noted in Chapter 7 is thus simultaneously a battle for control over environmental resources and a struggle over ideas concerning the best way in which to use and manage those resources. However, to appreciate fully this dual struggle is to understand the ways in which multilateral institutions like the FAO have provided states, businesses and other multilateral institutions with a particular vision of 'appropriate' environmental management based on scientific models and empirical research that has underpinned the development quest of these actors.

The best known example of this 'technical' support role relates to the provision of knowledge and advice surrounding the so-called Green Revolution technologies. These technologies were disseminated widely in the Third World beginning in the 1960s, and the combination of 'high-yielding' seeds and 'off-farm' inputs such as chemical fertilisers and pesticides was responsible for greatly enhanced crop yields in subsequent years (Pretty, 1995). Much of the actual research associated with such technologies was conducted through an international network of research stations coordinated by the Consultative Group on International Agricul-tural Research (a funding group comprising First World state and business interests plus the World Bank), but the FAO has been actively involved in the promotion of the ensuing 'products' (Richards, 1985). The FAO's promotion of Green Revolution technologies was coupled to a massive push to encourage Third World states to export agricultural cash crops as part of a broader assumption that the extent of modernisation was somehow related to the degree of integration in the global capitalist economy. In doing so, this organisation insisted that 'greater access to international markets would yield important benefits...[it] would massively increase the market in sugar, vegetable oils, tobacco, pulses, topical beverages, and forest products' (FAO cited in Goldsmith and Hildyard, 1991: 87).

The FAO campaign on behalf of Green Revolution technologies coincided with the postwar interest of Third World state leaders in the promotion of rapid economic development and industrialisation (see Chapter 3). Accordingly, it was highly successful, and Third World agrarian production became increasingly influenced by the new technologies (Pretty, 1995). In India, for example, Shiva (1991b) notes how Green Revolution technologies dominated the country's agricultural policies in the 1960s, such that, by 1968, one-tenth of wheat planted in the country was already of Green Revolution stock. Likewise, the use of pesticides – an essential component in the new 'package' – increased dramatically across the Third World: between 1972 and 1985 imports of pesticides increased by 201 per cent in Asia, 95 per cent in Africa and 48 per cent in Latin America (Dinham, 1991: 61). As a result of these 'miracle' technologies, between 1950 and 1985 'world grain output increased two and one-half times, growing at 3% a year' (Goldsmith and Hildyard, 1991: 82). The FAO sees in this record ample evidence of the validity of its strong support for a technical approach to Third World agriculture based on First World science:

> The outstanding fact in food and agriculture is that the past 25 years have brought a better-fed world despite an increase of 1.8 billion in world population. Earlier fears of chronic food shortages over much of the World proved unfounded.
>
> (FAO in Goldsmith and Hildyard, 1991: 82)

Yet the suggestion that FAO-sponsored Green Revolution technologies have been an unmitigated blessing for the Third World has been challenged sharply by political ecologists (among others). Looking at the promotion of these technologies, for example, Sesmou notes that although this process

> has arguably been a technical success, considerably increasing yields, it has proved an environmental, social and human disaster. Poor and small farmers have been systematically marginalised, the environment has been degraded, genetic diversity has been drastically eroded and the dependence of the South on the North has been increased.
>
> (Sesmou, 1991: 47)

The association of the spread of Green Revolution technologies with a growing shift to monocultural cash-crop production has been noted in the literature as being linked, in turn, to intensifying environmental degradation in the production areas (e.g. Franke and Chasin, 1980; Blaikie, 1985; Shiva, 1991b; Madeley, 1994; Pretty, 1995). In India, for example, Shiva (1991b) notes how this process has dramatically reduced diversity in agricultural production, and has been linked to the depletion of the soil's nutrients. An integral part of the 'success' of the Green Revolution has also been the massive application of chemical fertilisers and pesticides by farmers using the new high-yielding varieties of seeds. However, the use of chemical fertilisers and pesticides has had considerable adverse impacts on local ecological conditions wherever they have been used, and the pollution of rivers and soils has posed a growing health hazard to grassroots actors (Dinham, 1991). Health problems have been especially notable among the many poor farmers who have applied these chemicals in the fields (Bull, 1982).

Further, evidence has tended to suggest that the advent of Green Revolution technologies has been associated with a growing polarisation of rich and poor farmers in many parts of the Third World. Poorer farmers often find it very difficult to afford the needed inputs, including not only fertilisers and pesticides but also the seeds, which need regular replacement unlike seeds traditionally used (Pretty, 1995). As such, many farmers must borrow money to pay for the needed inputs, but escalating indebtedness has been the all too frequent result, as the anticipated benefits of using the new technologies have often failed to match the greatly increased costs of the production process itself. The introduction of these technologies has been associated in many areas with the growing affluence of wealthier farmers, often at the expense of poorer farmers for whom the Green Revolution has been mainly about growing political, economic and ecological marginalisation (see Chapter 2). Thus, the widely held assumption at technical organisations like the FAO that Green Revolution technologies would improve the lot of poor grassroots actors was, to say the least, 'a trifle heroic' (Blaikie, 1985: 18).

At a general level, the Green Revolution has been linked to the increased dependency of many Third World countries on the First World. On the one hand, the Green Revolution has been all too successful in its underlying aim of promoting increased yields frequently linked to the elaboration of an export-oriented cash-crop-based economy. Thus, while the FAO rightly declares that world food production has increased over the past fifty years, much of this increase has been concentrated in just a few crops grown mainly for export: coffee, cotton, wheat, groundnuts, and so on (Watts, 1983a; Rush, 1991; Kendie, 1995). If such production has been good news mainly for First World businesses and consumers, it has often been bad news for many Third World grassroots actors – even for those countries whose economic well-being has been tied to the market fortunes of one or two primary products (see Chapter 3).

On the other hand, the Green Revolution has resulted in a growing dependency of the Third World on First World agro-chemical industries who control the new 'miracle' technologies. As Chapter 5 notes, the quest to accumulate capital has led agro-chemical TNCs to develop 'integrated' packages (i.e. seeds, fertilisers, pesticides) in order to maximise profits on sales to Third World farmers. These businesses have been highly appreciative of the efforts of 'technical' organisations like the FAO to sell these integrated packages in the Third World. Indeed, the FAO has enjoyed generally close links to the agro-chemical industry, and such links are perpetuated notably through 'the industry's presence at FAO workshops, industry observers at expert panels and committees, and through common goals such as pesticide registration and standards' (Dinham, 1991: 62).

Yet if many Third World countries have become increasingly dependent on the First World, Third World states have nonetheless enjoyed selected benefits through their ability to control the flow of FAO funds into their countries. While nowhere near as significant as the funds provided by IFIs led by the World Bank (see below), these funds have still been an important means by which Third World states have sought to strengthen their agricultural management capacities. FAO funds have thus been highly useful for states keen to enhance their power over grassroots actors, especially in those states where control over rural populations has been relatively weak (see Chapter 3). In this manner, 'technical assistance' has been an invaluable tool to assert essentially political goals. At a more prosaic level, FAO funds have been a boon to cash-starved state agencies. Indeed, using examples from Africa, Sesmou (1991: 52) argues that the impact of the FAO's agricultural development projects at the local scale is 'of marginal importance: their real role is to support the budget of the Ministry of Agriculture by providing salaries and equipment'. In the process, those grassroots actors for whom these FAO projects are formally intended often see few, if any, of the benefits that the FAO claim to be providing through their initiatives.

The preceding discussion has referred only to FAO activities in relation to agriculture, but has nonetheless raised questions about the non-political status of this organisation. It has further cast doubt on FAO claims to be promoting social justice and environmental conservation goals through its activities. Looking at this particular aspect, McCormick (1995: 33) notes that, given its goal to promote enhanced long-term world food productivity, the FAO 'could not avoid making resource conservation a central item on its agenda'. Indeed, Article 1 of the FAO's constitution specifies that the main objectives of the organisation are 'the conservation of natural resources and the adoption of improved methods of agricultural production' (McCormick, 1995: 33). However, as we have seen, the FAO's record has been largely one of promoting the latter at the expense of the former.

As with other multilateral institutions (see below), the FAO is increasingly having to justify its activities, and their social and environmental effects. For example, the FAO has recently adopted a 'sustainable agriculture and rural development' programme (SARD) as part of an effort 'to insert its policies into Agenda 21' (Hildyard, 1991: 239). However, this programme has been condemned as being nothing more than a repackaged version of its traditional 'productivist' stance. Thus, SARD continues to emphasise intensive agricultural production, including a continued reliance on 'off-farm inputs' (i.e. fertilisers, pesticides). Indeed, the views contained in SARD are very similar to those set out in an earlier policy document published in 1987, but with an added gloss on 'sustainability' to mollify critics of the FAO. Thus, instead of arguing that intensification is needed to 'transform agriculture into a dynamic, productive sector', SARD suggests that such intensification is also a valuable means to 'protect' the environment, to 'enable' land reform, to 'secure' biodiversity, and to 'relieve rural poverty' (Hildyard, 1991: 241). Intensive agriculture and sustainable environmental management are 'essentially compatible' in that the former is required 'to avoid encroachment of agriculture on areas which should be protected or used otherwise, and to relieve production pressures on fragile marginal arable lands and grazing lands' (FAO cited in Hildyard, 1991: 241). With this continued concentration on agricultural intensification, it would appear likely that the SARD programme will not result in any appreciable changes to the FAO's policies – that is, policies that have notably contributed to environmental degradation and associated social marginalisation and poverty in the past. The role of the FAO (and technical organisations in general) nonetheless pales in comparison with the role of the World Bank and other IFIs in Third World environmental change.

INTERNATIONAL FINANCIAL INSTITUTIONS

Whilst United Nations technical organisations like the FAO have had a notable impact on Third World peoples and environments, it is the international financial institutions (IFIs), led by the World Bank, which have had the greatest impact overall, and it is these latter institutions which have garnered the lion's share of the public criticism directed at multilateral institutions in recent years. This section examines the contribution of IFIs to Third World environmental change and conflict, focusing in particular on the role of the much-maligned World Bank in these inter-related processes.

The World Bank has been singled out for extensive study by political ecologists due to the sheer size and importance of this institution's role in environment and development issues in the Third World (Adams, 1991; George and Sabelli, 1994; Rich, 1994). The World Bank stands as the Third World's single largest creditor, and of a total Third World debt in 1990 of $1.3 trillion it was owed $182 billion (Piddington, 1992; Rich, 1994). Yet the importance of the World Bank extends well beyond this fact since this actor serves as the lead agency (in conjunction with the IMF, see below) in the provision of First World private and public lending to the Third World. In other words, until a Third World country receives the World Bank's 'seal of approval', it has virtually no chance of receiving any loans from the First World at all. This ability to regulate the flow of financial resources from First World states and banks to Third World states is the source of considerable power for the World Bank, especially in a context of massive and pervasive Third World indebtedness (George, 1988).

However, the power and importance of the World Bank reside not only in its role as the Third World's banker but also, and above all, in its ability to influence the development trajectory of most Third World countries over the five decades of its existence. During this brief period, it has developed a 'secular empire' (George and Sabelli, 1994) of unparalleled proportions, which has been a major contributor to the Third World's environmental crisis. Founded in 1944 as the International Bank for Reconstruction and Development (IBRD), the World Bank, as it later became known, was created initially to facilitate Europe's postwar recovery, but soon shifted its focus to the provision of 'development' assistance to the Third World. A creature of the First World, this institution has been controlled throughout its existence by the United States and other leading First World countries through an executive board in which voting power is related directly to the amount of capital that a country has invested in the Bank (Piddington, 1992).

The rise of the World Bank to become the most powerful development agency in the world did not occur naturally, but rather reflected a skilful campaign by the Bank in which it created a demand for its loans and services through a process of institution building and technical training (Rich, 1994). Its favoured tactic in this regard was to encourage Third World leaders to create development 'parastatals' – autonomous state agencies whose primary function was to promote economic development, notably through the building of infrastructure. Spurred on by this actor, during the 1950s and 1960s most Third World states created state development boards, financial institutions and electricity-generating companies, whose management was not accountable to the people of the Third World, even though the massive debts that these parastatals quickly incurred were ultimately the liability of those people. In this manner, the World Bank and these newly created parastatals engaged in a mutually rewarding pattern of interaction: the latter borrowed heavily from the former as part of their institutional growth, which created, in turn, a strong continuous demand for World Bank loans, thereby justifying further World Bank expansion, and so on. The Bank's role as a leading development institution was further strengthened in 1960 with the creation of the International Development Association (IDA) which was to make 'soft loans to the World's poorest countries unable to afford IBRD terms' (George and Sabelli, 1994: 11). Ostensibly a non-political actor, the World Bank has been, in practice, highly influential politically through its 'unique and unprecedented mechanisms for continual intervention in the internal affairs of borrowing countries' (Rich, 1994: 10).

The World Bank has used that influence to promote a vision of development predicated on large-scale projects that found a receptive audience among Third World leaders and industrialists impatient to share in the fruits of economic 'modernisation' (Peet, 1991; Hettne, 1995).

However, as this book has already suggested, postwar economic develop-ment has been a central factor in producing the social inequities and environmental degradation that are characteristic features of the Third World's politicised environment. For a long time the World Bank was seemingly oblivious as an institution to this dark side of its lending policy, caught up, as it was, in an institutionally-rewarding process of funding ever more grandiose development projects relating to logging, mining, transmigration, cash-crop production, cattle ranching and dam construction in most parts of the non-communist Third World (Hecht and Cockburn, 1989; Hurst, 1990; Adams, 1991). Initial criticism of the adverse social and environmental effects of Bank lending certainly led in the 1970s to a wider remit encompassing questions pertaining to basic needs, rural development and environmental quality. Big development projects none-theless remained the top priority, and new social and environmental concerns did not alter the basic thrust of World Bank lending policy. For example, despite rhetoric about its 'environmental programme', this institution employed only six staff on environmental matters out of a total staff of 6,000 in 1983, and environmental reviews of planned projects scarcely figured in Bank thinking about project viability (Le Prestre, 1989; Williams, 1993; Rich, 1994). The prevailing view within the Bank was that 'selective' social and environmental degradation was a small price to pay for Third World development.

As political ecologists have documented, that price was in reality neither small nor equitably distributed in society. Indeed, with the help of the World Bank and other IFIs (see below), states often acting in conjunction with local and transnational businesses have transformed the topography of the Third World's politicised environment. Thus, in a manner reminiscent of colonial times (see Chapter 3), IFI-sponsored 'progress' was measured in terms of trees felled, valleys flooded, minerals extracted and acreage dedicated to cash crops or cattle ranching. The flip-side of such 'progress', as Chapter 7 shows, was ruined livelihoods and degraded environments as the environmental resources of poor grassroots actors were despoiled by loggers, engineers, mining companies, cattle ranchers, agri-businesses or land-hungry migrants encouraged by state and World Bank officials (Franke and Chasin, 1980; Hecht, 1985; Hall, 1989; Adams, 1990; Hurst, 1990; Cummings, 1990). It is true that this process cannot be attributed solely to the World Bank and IFIs as powerful state and business interests were also behind these various activities, and some of these activities (e.g. transmigration in Indonesia) predated IFI involvement. However, the role of these multilateral institutions in this process was to enable other actors to pursue socially and environmentally destructive practices on a scale hitherto impossible. Two examples will serve to illustrate the adverse social and environmental effects of World Bank-sponsored development projects.

The first example relates to a large-scale energy project in India that was designed to increase the country's energy supplies so as to meet the rapidly growing national demand. In the late 1970s, the World Bank supported plans by the Indian state to develop a super thermal power plant, an open-pit coal mine and associated electric transmission lines in the Singrauli region of north-central India. This project provided much-needed energy for national industrial growth, but only did so at an immense social and environmental cost. As Rich observes of the area today,

> Enormous quantities of coal dust and ash pollute the air. Eight cement plants and thousands of stone crushers release over a thousand tons of cement and rock dust daily into the atmosphere. The five thermal coal plants belch out huge quantities of sulphur dioxide – none are equipped with scrubbers – and emit over 1,650 pounds of mercury a year into the atmosphere. The Riband dam reservoir and the land around it are poisoned by dangerous concentrations of mercury, fluorine, and chromium, and as a result the crops and fish that the half-million people in the Singrauli area consume are in many cases unfit for human consumption. The productivity of the land has been destroyed, the once drinkable ground water is contaminated.
>
> (Rich, 1994: 40)

As this description indicates, social immiseration has accompanied environmental degradation as the project has resulted in the displacement of thousands of poor grassroots actors and the disruption of untold thousands of livelihoods as a result of the destruction of vast areas of agricultural land, and the poisoning of local fish and water supplies. Thus,

> the situation of many of the hundreds of thousands of the local inhabitants has degenerated from traditional poverty in what was a society based on subsistence agriculture thirty years ago to absolute destitution. Each time they were forced to move without compensation or rehabilitation they became poorer.
>
> (Rich, 1994: 40)

A second example of a World Bank-sponsored development project, inimical to both the interests of poor local grassroots actors and the environment, concerns Brazil's Greater Carajas Programme (PGC) discussed briefly in Chapter 3. This project, which the Bank funded to the tune of $305 million in the early 1980s, is centred on an iron-ore mine, linked pig-iron smelters and a 900 kilometre railway connecting the mine to a deep-sea port located at Sao Luis. This gigantic 900,000 square kilometre project also encompasses agricultural, livestock and forestry activities as part of a major attempt to 'develop' eastern Amazonia. Yet the social and environmental costs associated with the PGC have been equally spectacular, notwithstanding the social and environmental provisions incorporated in the loan provisions. Vast areas of forest needed to be cleared to make way for the mine and smelter sites, the railway and associated settlements. The most significant cost was linked to the demand for charcoal for use in the smelters, which was estimated to require the destruction of the equivalent of 1,500 square kilometres of rainforest per year (Anderson, 1990). A further adverse environmental impact was associated with the influx of farmers and entrepreneurs attracted to this new Amazonian frontier who accelerated land clearance in the area, thereby repeating a process played out in the Amazon generally (Hecht and Cockburn, 1989; Schmink and Wood, 1992). The social impact on the region's one million farmers and indigenous peoples has also been severe. Thus, the transformation of the Carajas region has been marked by the widespread displacement of local grassroots actors as companies linked to the PGC, as well as opportunistic migrants, 'grab' land from long-standing occupants (Hall, 1989).

As the Brazilian example illustrates, the World Bank persisted with a policy of funding large-scale development projects into the 1980s even though by then criticism of this policy had already begun to mount, especially criticism from ENGOs located in both the First and Third Worlds (see below). However, the 1980s also marked a dramatic new role for the World Bank in relation to Third World environmental change and conflict as a result of the onset of the Third World debt crisis in 1982. This crisis has been extensively reviewed in the literature, but what is important here is the effect of World Bank (and IMF) 'structural adjustment' lending policies on the policies and practices of Third World states and, by extension, on environmental change and conflict in the Third World (George, 1988, 1992; Adams, 1991). The development of structural adjustment policies (SAPs) was an attempt by First World states and banks, acting through the World Bank and the IMF, to stave off the potential collapse of the global capitalist system in the event of widespread loan defaults by heavily indebted Third World states. As noted in Chapter 3, Third World states were usually required under these SAPs to curtail government spending (hitting especially hard the budgets of fledgling environmental agencies) and to promote a pro-market and export-oriented development strategy that typically boiled down to enhanced exploitation of natural resources. Thus George (1988: 156) notes that 'there are two debt/environment connections. The first is borrowing to finance ecologically destructive projects. The second is paying for them ... by cashing in natural resources.'

The Third World debt crisis also marked the rise of the IMF as a major actor involved in Third World development and environment issues. In contrast to the World Bank, the role of the IMF since its creation in 1944 has been to ensure the regulation of member countries' balance of payments, and thus to ensure that they 'pay their international debts' (Adams, 1991: 76). As a result, during the 'boom years' of the 1950s, 1960s and 1970s, the IMF played a minimal role in Third World affairs but, with the onset of the debt crisis, this institution came into its own as the First World's 'financial police officer'. Acting in conjunction with the World Bank, the IMF required Third World states to submit to severe austerity measures involving the 'structural adjustment' of state policies and economic activities before loans were disbursed. The latter has been the signal for other First World lenders to renew loans to Third World states (Adams, 1991).

Profound social and environmental effects have been associated with IMF and World Bank structural adjustment policies in the Third World. At a general level, SAPs have deepened the political and economic dependency of most of the Third World on the First World (George, 1992). It is at the local scale, though, that these effects have been most acutely felt, with the overall result being intensified social poverty and environmental degradation (Reed, 1992). Environmentally destructive practices perpetrated in the post-Second World War era by states and businesses (often assisted by multilateral institutions) have been extended to meet national debt repayment schedules set by the World Bank and the IMF. Indeed, the export-led growth model on which SAPs have been based is 'a purely extractive one involving more the "mining" than the management – much less the conservation – of resources' (George, 1992: 2–3). In the case of Ghana, for example, the introduction of SAPs in the 1980s was praised by the World Bank as an example of 'model behaviour', but this process involved 'expanding mineral production and logging at a rate that could be characterised as pillage' (Rich, 1994: 188). The expansion of commercial logging has resulted in the reduction of the country's forest area to 25 per cent of its original size, with timber exports increasing from $16 million in 1983 to $99 million in 1988 (Rich, 1994). In Ghana, as in many other Third World countries, SAPs have also contributed to a push to intensify export-oriented cash-crop production with associated instances of soil nutrient depletion and grassroots actor marginalisation (Reed, 1992; Kendie, 1995). In Senegal, groundnut production was expanded during the 1980s to one million tonnes of groundnuts per annum but, as a consequence, the country's soils are 'so depleted by groundnut production that today they can produce nowhere near that amount' (George, 1992: 3). Yet, this process (as elsewhere in the Third World) has had a ripple effect in so far as the promotion of industrial agriculture has resulted in the displacement of poor grassroots actors whose land has been appropriated by more powerful state and agri-business actors (Davidson et al., 1992). In other cases, these farmers themselves have been the agents of environmental degradation as they have been forced into unsustainable land-management practices through contractual obligations to state and business corporations. Such 'contract farming' is increasingly prevalent in the Third World today as TNCs and other large corporations seek to control ever more closely the agrarian production process (Little and Watts, 1994).

World Bank- and IMF-sponsored SAPs have also been linked to increasing health and pollution problems arising from unregulated industrialisation. In Mexico, for example, George (1992) and Coote (1995) note how the state has encouraged the relocation of American manufacturers from the United States to the Mexican side of the joint border partly as a response to Mexico's massive national debt. Yet within this industrialising *maquiladora* zone', 'there are few environmental regulations... and virtually none that Mexico can now afford to enforce' (George, 1992: 25). The social and environmental impacts of such development include chemical leakages, toxic fume emissions, and a range of illnesses experienced by local workers and residents (Coote, 1995; see also Chapter 5).

As these examples illustrate, the World Bank's role in Third World environmental change and conflict combines the promotion of traditional large-scale development projects, and, since the early 1980s, the imposition of structural adjustment packages (in conjunction with the IMF) on heavily indebted Third World states. The World Bank was heavily criticised by ENGOs and grassroots organisation leaders on both counts during the 1980s, such that by the end of that decade, the Bank was keen to cultivate a 'green' image to stave off growing opposition to its development policies. The role of ENGOs and grassroots organisations is discussed in Chapters 6 and 7, but here it is important to note that the World Bank has been an important target of these actors precisely because of the perceived political importance of that institution in Third World environment and development issues.

Although the World Bank was attacked over its involvement in various projects, it was the ENGO campaign against the Bank's support for Brazil's massive resettlement and integrated rural development project (POLO-NOROESTE) in Rondonia in the Amazon in the early 1980s that ultimately marked a watershed in World Bank operations (Hecht and Cockburn, 1989; Bramble and Porter, 1992; Rich, 1994). This project, as with many other large projects sponsored by the World Bank, was designed to develop a hitherto 'peripheral' region but, as was so frequently the case, it was the cause of pervasive environmental degradation and social conflict and dislocation. Protests by ENGOs directed to the World Bank achieved little, but a well-publicised campaign conducted in the United States Congress by US ENGOs led to a highly significant ENGO victory in March 1985, when the combination of ENGO and Congressional pressure forced the Bank to halt remaining loan disbursements in support of that project 'pending the preparation and carrying out of emergency environmental and Indian lands protection measures by the Brazilian government' (Rich, 1994: 126).

Defeat over the POLONOROESTE project was a major blow to the World Bank. The Bank was used to getting its own way over environment and development matters, but this defeat signalled that there were definite limits to its power.

For the first time, the Bank was forced to account to outside NGOs and a parliamentarian of a member country for the environmental and social impacts of a lending program, and for the first time a public international financial institution had halted disbursement on a loan for environmental reasons.

(Rich, 1994: 127)

The POLONOROESTE defeat also led to a publicised 're-think' on the Bank's part as to its position on environmental matters. In 1987, World Bank President Barber Conable conceded that there had been 'mistakes in previous bank environmental policy' (cited in Williams, 1993: 130), and set out a new commitment on the part of this institution to incorporate systematically environmental considerations in future lending policy (Piddington, 1992; Rich, 1994). In addition, the Bank was to fund projects geared specifically to addressing environmental problems such as deforestation, and the means by which it was going to maintain this environmental commitment was through a newly created central Environmental Department (ED) acting in tandem with environmental units located in the World Bank's four main regional divisions (Le Prestre, 1989). In an intriguing attempt to win over its erstwhile opponents, the World Bank even sought input from ENGOs over Bank operations (Rich, 1994).

World Bank environmentalism after 1987 also reflected broader institutional calculations linked to the growing popularity of 'sustainable development' thinking in international policy-making circles following the publication of the Brundtland Commission report (World Commission on Environment and Development, 1987; see also Introduction). Just as TNCs found the message contained in the report congenial to continued capital accumulation (see Chapter 5), so too the World Bank became enamoured with a concept that suggested that the integration of economic development and environmental conservation was not only feasible, but also imperative, if the Third World's environment and development problems were ever to be resolved. Thus, the late 1980s and early 1990s witnessed a concerted effort by the World Bank to assert itself as a world authority on environmental issues (to complement its existing development preeminence). In doing so, this institution attributed the Third World's environmental crisis largely to the twin problems of poverty and population growth – although it conceded that 'rapid and uncontrolled economic growth' was also partly to blame (Williams, 1993: 132; see also World Bank, 1992).

The 'greening' of the World Bank since 1987 has been expressed through an attempt to develop 'many ways of integrating environmental issues into development policy-making' (Williams, 1993: 131). At an aggregate level, this process was reflected in growing attention to environmental questions in Bank projects. Thus, while the number of projects with a 'significant' environmental component was negligible prior to the mid-1980s, by the end of that decade such projects represented between one-third and one-half of total Bank projects (Williams, 1993: 131). Further, environmental lending also expanded rapidly totalling some $1.6 billion (about 7 per cent of total Bank lending) in 1991 (Piddington, 1992). Such lending was designed to tackle diverse environmental problems, but also aimed to strengthen environmental agencies in the Third World. Brazil, for example, received $117 million in the 1990s to strengthen its environmental agencies (Rich, 1994: 148). The World Bank's environmental staff were also empowered in 1990 to 'undertake full environmental impact assessments on every project with the potential for substantial environmental effects' (Williams, 1993: 131).

Finally, and in keeping with the Brundtland Commission's enjoinder to official agencies to pay greater attention to the wishes and needs of the poor, the World Bank has sought to acquire and incorporate input from poor and marginal grassroots actors as an integral part of project evaluation. As Davidson et al. note,

> one important innovation is that Bank staff are now expected to seek the views of the poor who may be adversely affected by a proposed project. The Bank now acknowledges that project success often depends on effective local participation throughout the project cycle.
>
> (Davidson et al., 1992: 180)

Cooperation with grassroots organisations which voice the concerns and interests of the poor has been essential to this task (Piddington, 1992).

However, the apparent 'greening' of the World Bank has been subject to intense scrutiny and criticism by activists and scholars alike. To begin with, critics have argued that this process has more to do with appeasing the United States government and powerful US ENGOs than it has to do with a fundamental reevaluation of the World Bank's role in the development business (Williams, 1993). Some critics even allege that the greening process is nothing more than a façade for a business-as-usual approach in keeping with the Bank's original development remit. In the quest to promote economic development and to maintain, if not enhance, its institutional power as the world's leading development agency, social and environmental concerns will always play second fiddle to the Bank's main goal of promoting economic development (Davidson et al., 1992; George and Sabelli, 1994). Indeed, and despite its new-found environmental 'sensitivity', the World Bank has continued to subscribe to development mega-projects linked to adverse social and environmental impacts, and has only desisted from such support when confronted with massive and well-publicised environmental protests.

The case of the Narmada Dam project in central India is instructive in this regard. Following an agreement between the Indian state and the World Bank, construction of a series of large, medium and small dams along the Narmada river and its tributaries began in 1987, but each stage of the project's development was dogged by popular protests and resistance (see

Chapter 7). Notwithstanding clear evidence provided by the Bank's own enquiry that the project would result in widespread social hardship and environmental degradation, it persisted in its support of the proposed dam complex, only backing out of the deal in March 1993 when the political campaign to 'stop the Narmada' threatened to end in a repeat of the POLONOROESTE fiasco (Rich, 1994).

An even more severe criticism of the idea of a 'green' World Bank is the argument that this new image clashes with the reality of that institution's role as a leading promoter of structural adjustment policies in the Third World (Reed, 1992). Chapter 3 noted how Third World states have been expected to adopt green policies and practices in a context characterised by the retrenchment of state services and the need to boost export earnings to repay debts. Here, however, it is important to note that 'a clear contradiction emerged between the Bank's efforts to deal with the macroeconomic crisis of debt and adjustment, and its purported goals of poverty alleviation and increased attention to environmental concerns' (Rich, 1994: 186; see also Dore, 1996). The initial response of the World Bank to this apparent contradiction was to claim that there was really no contradiction at all in the long term. It conceded that there were 'some short-term adverse effects on the poor, but that its adjustment policies were necessary to ensure "long-term sustainable growth", which, it claimed, would benefit all' (Rich, 1994: 189). However, in 1989, the Bank began to make some attempt to mitigate the adverse environmental and social impacts associated with structural adjustment lending, through, for example, the integration of environmental assessments into structural adjustment lending policy (Williams, 1993; Mikesell and Williams, 1992).

Critics allege that these efforts have been ineffectual, and 'probably have served more of a political purpose in giving adjustment the appearance of a human face, rather than a genuine compensatory purpose' (Oxfam cited in Rich, 1994: 189). As a 'straightforward' financial institution without any social or environmental policy pretensions like the World Bank, the IMF has been largely removed from this political fray, although the early 1990s witnessed a limited acknowledgement by this institution that it must at least consider environmental issues in its activities (Mikesell and Williams, 1992). As the world's leading development agency, the World Bank has been much less able to wriggle out of the difficulties occasioned by its contradictory mandates.

Even the World Bank's initial forays into the realm of environmental policy lending have met with considerable criticism. The argument is that such lending has done very little to resolve the problems of environmental degradation that it has set out to deal with, but rather is merely a new way for the World Bank to maintain its predominance in Third World development planning (George and Sabelli, 1994; Rich, 1994). More damning still is the claim that, in some cases of environmental lending by the World Bank, environmental degradation has increased rather than decreased.

The case of the Tropical Forestry Action Plan (TFAP) is apposite here. This major project has included the World Bank, the United Nations Development Program, the World Resources Institute and the FAO, with the latter being the lead agency involved, given its prominent 'expertise' in forestry (Marshall, 1991). The TFAP was established in 1985 to coordinate the flow of an estimated $8 billion in development assistance from the First World to the Third World, and was designed to assist participating Third World states in efforts to halt tropical deforestation within their borders (Food and Agriculture Organisation, 1987). However, the project was the subject of intense criticism from the start by many ENGOs and grassroots organisations who objected to the secretive 'top-down' nature of the project, and the concomitant failure to consult with those grassroots actors most dependent on the tropical forests (Colchester and Lohmann, 1990). Indeed, it was claimed that since the TFAP was essentially 'a plan by foresters for promoting forestry', the project was really a means by which powerful state, business and multilateral institutions sought to increase large-scale commercial logging even though there was no 'convincing evidence' that such logging 'could be made both sustainable and economically viable' (Colchester and Lohmann, 1990: 84). Above all, the TFAP, as expressed through specific national plans, failed to 'give priority to the poor, to women or to marginalised social groups, notably indigenous people', and, as such, reflected a 'business-as-usual' approach to environmental management (Colchester and Lohmann, 1990: 85).

Undeterred by such criticism, the World Bank has moved in the 1990s to consolidate its position as the lead agency in 'sustainable development', assuming the mantle of 'global environmental manager' through its control over 'green' aid flows between the First and Third Worlds. The primary way in which it has done so has been through management of the newly created Global Environmental Facility (GEF). Arising from a suggestion made by the French at the 1989 G-7 summit, the GEF was established under the World Bank's control in November 1990 to provide a global mechanism to finance conservation projects and strategies in the fields of energy efficiency, forestry, biodiversity preservation, water pollution and protection of the ozone layer (Gan, 1993; Chatterjee and Finger, 1994). By the end of 1992, seventy projects, worth $584 million in total, had been approved through the GEF (Chatterjee and Finger, 1994). The GEF received a further boost in June 1992 when this mechanism was established at the Rio Earth Summit as the primary means by which the First World would financially assist Third World states in their efforts to implement sustainable development policies. To this end, the GEF's core funding was set to go up to as much as $4.2 billion in order to meet its rapidly expanding functions (Chatterjee and Finger, 1994).

The World Bank's bid to become a 'global environmental manager' has encountered much criticism from many ENGOs and Third World states. The fact that the GEF was designed behind closed doors by the Bank suggested to these actors that this institution remained as unaccountable and insensitive to Third World needs as ever (Tickell and Hildyard, 1992; Rich, 1994). Further, evidence suggested that the Bank was simply using the GEF as a new means to promote its favoured development goal, and that, as in the case of the TFAP, the GEF was not necessarily about environmental conservation at all.

Critics cite the example of the GEF-funded Congo Wildlands Protection and Management Project developed in 1991. The project was devised exclusively by the World Bank in consultation with officials in the Congo, and aimed 'to protect bio-diversity in the rich, untouched Nouabele rainforest in the North of the Republic of the Congo' (Rich, 1994: 178). Yet this GEF project was soon overshadowed by leaked reports that the Bank was simultaneously planning to finance separately commercial logging in the country through funds designed to facilitate the transport and export of logs (Rich, 1994). As a consequence of this, and similar situations, it is not surprising that 'the GEF is being viewed by dam-builders, loggers and the like as a new source of subsidies – provided that they can use their public relations skills to portray their projects as beneficial to the global environment' (Tickell and Hildyard, 1992: 83).

The World Bank's role in relation to the Third World's politicised environment in the 1990s would thus appear to smack of 'old wine in a new bottle'. The social and environmental impact of the Bank during the postwar era has derived from its role as the leading development agency in the Third World. This institution seeks to perpetuate that role in the 1990s through allegiance to the concept of sustainable development, and through its necessary role in the promotion of that concept throughout the Third World.

A similar pattern of institutional development has characterised the experience of the regional IFIs created in the aftermath of the Second World War with briefs comparable to that of the World Bank but in the specific contexts of Latin America, Africa and Asia. The Inter-American Development Bank (IADB), the Asian Development Bank (ADB) and the African Development Bank (AfDB) adopted roles very similar to that of the World Bank in so far as they privileged development goals over social and environmental considerations (Adams, 1991; Mikesell and Williams, 1992). Indeed, Adams (1991: 70) argues that the World Bank and the three regional Banks together 'all virtually have the same articles of agreement, structures and lending patterns, leading precisely to the same dismal results'.

In many cases, the regional IFIs cooperated directly with the World Bank in providing loans to big development projects. In Indonesia, for example, the ADB was a major supporter of that country's transmigration programme in which millions of poor landless Javanese farmers were persuaded over the years to move from 'overpopulated' Java to the 'underpopulated' Outer Islands (Hurst, 1990). In Brazil, the IADB pursued a similar funding strategy with regard to that state's Amazonian transmigration scheme (POLONOROESTE) (Hecht and Cockburn, 1989). In both cases, regional bank funding was designed to coordinate with separate funds provided by the World Bank. However, just as the latter has been criticised for its support of large-scale transmigration schemes, so too have the regional banks encountered the ire of ENGOs and grassroots organisations over their lending programmes (Hurst, 1990; Bramble and Porter, 1992).

In a similar vein, the greening of the regional IFIs has paralleled that of the World Bank in the late 1980s and 1990s. Mikesell and Williams (1992: 19) note that, 'as in the case of the World Bank, regional Banks became aware of the environmental consequences of MDB [Multilateral Development Bank] supported projects long after environmentalists began calling the attention of the international community to the serious problems in Third World countries'. To take the case of the IADB, for example, this bank did not establish its Environmental Protection Division (EPD) until 1989, and only then after mounting protests (notably by US ENGOs) in the mid- to late 1980s over the IADB's promotion of environmentally destructive projects in the Amazon (Bramble and Porter, 1992; Schmink and Wood, 1992). The EPD's role was comparable to that of the World Bank's Environmental Department in that it was created with a brief 'to evaluate the environmental impact of projects and to recommend remedial action' (IADB cited in Mikesell and Williams, 1992: 20). However, once again, the EPD's role was peripheral to the 'real' business of the IADB – namely, the promotion of development projects. The IADB thus remains a target of ENGO and grassroots organisation criticisms along with the World Bank and its counterpart agencies in Africa and Asia, which have been subject to comparable attacks for their parochial pursuit of 'development' (Rich, 1994; McD.Beckles, 1996).

The role of the IFIs in the Third World's environmental crisis has been to promote a vision of development essentially incompatible with the requirements of social equity or environmental conservation in the many places in which the effects of their lending policies have been felt. The World Bank as the leading development agency has been especially central in this regard and, notwithstanding its efforts to cultivate a green image, there is little indication today that this pivotal actor has substantively modified its 'development-first' approach.

Thus, the overall impression given in this chapter so far is that the contribution of multilateral institutions has mainly been a negative one linked to social and environmental degradation. In the next section, we briefly consider why this has been the case, as well as query whether there are opportunities within the contemporary United Nations system for an alternative approach.

HANDMAIDENS OR POWER-BROKERS?

The role of multilateral institutions in the Third World's environmental crisis has been an indirect, but nonetheless powerful one. These actors have sought to maintain a non-political status in the pursuit of 'technical' questions pertaining mainly to the provision of knowledge and finance to Third World states anxious to 'modernise'. However, as the advice and funds that they provided contributed mainly to social poverty and environmental degradation, mounting criticism from ENGOs and grassroots organisations meant that they could no longer avoid the political consequences of their actions.

In seeking to understand why multilateral institutions have been wedded for so long to development projects injurious to both people and environments in the Third World, it is important to emphasise two things. First, these multilateral institutions, and especially the IFIs, have been closely associated with, if not completely dependent on, the major First World states. These institutions have tended to adopt, not surprisingly, the interests and mores of those states but, in doing so, have almost inevitably distanced themselves from those poor grassroots actors in the Third World which many multilateral institutions are committed (at least in theory) to assisting. The point here is not that the main multilateral institutions have necessarily or always promoted First World interests, but rather that their overall outlook has been congruent with the prevailing views and interests of the First World. This situation is most evident with regard to the IFIs, which were created at the behest of the United States and other key First World states at the end of the Second World War, and which remain largely dependent on those countries for continued political and economic support. Thus, the policies of IFIs have conformed generally to those advocated by the First World – whether it be Keynesian-style interventionist strategies during the 1940s through to the 1970s or the neo-classical pro-market approaches adopted since the late 1970s and early 1980s. However, as the discussion of the FAO suggested, this 'policy congruence' extends to the FAO as well, as this organisation has sought to promote 'modernisation' in the natural resources sectors of Third World countries as part of the full integration of those countries into a First World-dominated global capitalist system (see Chapter 5).

Second, the major multilateral institutions have also promoted these development policies because they have shared a believe that 'development' is a process worth pursuing above all else, and that the Third World's social and environmental problems boil down, ultimately, to a 'lack of development'. As various scholars have shown, this fixation on development reflects a long history in which the First World has been critical of the economic and scientific standards of the Third World, and in which it has appointed itself as 'catalyst' of economic and scientific 'progress' in the Third

World (e.g. Adas, 1989; Sachs, 1992; Watts, 1993). Indeed, the very definition of the 'Third World' itself has been a reflection of this process (Said, 1978; Escobar, 1988). The role of the major multilateral institutions in the procedure has been to serve as intermediaries or midwives to development. However, in so far as such development is predicated on the elaboration of a global capitalist system seemingly at odds, by its very nature, with sustainable environmental management and local social justice (see Chapter 5), these institutions appear to be condemned to perpetuating practices that belie the formal mandate on which their activities are ostensibly based.

Given that formal mandate and the fact that not all multilateral institutions are completely beholden to the First World, what are the prospects of autonomy for these institutions working under the auspices of the United Nations, in terms of Third World environment and development issues? A useful way in which to approach this question is to consider the prospects of a coherent and unified policy on Third World environmental matters emerging from the United Nations system, a policy that does not simply privilege development issues like the key multilateral institutions discussed in this chapter.

In this regard, the experience of the United Nations Environment Programme (UNEP) is instructive. Founded in the wake of the 1972 Stockholm Conference, this agency could claim a certain representativeness derived from its history as a product of the world's first major international conference on the environment (McCormick, 1995; Momtaz, 1996). Although the UNEP is not a 'line agency' (like the FAO, for example), but rather an environmental policy coordinating body for all United Nations organisations, it has been well placed in principle to oversee the development of a coherent and integrated United Nations policy on the environment. That the headquarters of this agency is located in a Third World country (Kenya) suggests, if only symbolically, that Third World interests and concerns may receive priority in the United Nations system.

The history of the UNEP has nonetheless been a record of disappointment and frustration – disappointment in that it has never been granted the power to pursue its mandate of developing an integrated environmental approach within the United Nations; and frustration as it has been forced to watch helplessly as other United Nations organisations and IFIs have undermined its modest efforts to effect change on environmental issues. Indeed, as McCormick (1995: 139) notes, 'its job of persuading other agencies to execute programmes was hampered by the fact that it had few incentives to offer and no powers of enforcement'.

The failure of the UNEP to assert a powerful role for itself has been a particular blow to many Third World countries since this Third World-based agency was viewed as 'their own environmental agency' (McCormick, 1995: 138). Indeed, as Caldwell (1990: 75) notes, 'UNEP has provided a forum acceptable to the Third World countries for examining their mutual problems free from suspicion of solutions imposed by the First World'. Yet it is probably precisely this role for the UNEP that has scuppered its chances of becoming a leading agency within the United Nations system. There is thus little chance that First World states (led by the United States) will permit the UNEP the necessary funds and personnel to allow that agency to mount an effective campaign on behalf of the Third World in the United Nations system. Indeed, part of the explanation for the 'greening' of institutions such as the FAO or the World Bank may be that it enables these First World-dominated institutions to maintain their leading position in relation to institutions, such as the UNEP, that are seemingly more sympathetic to the Third World.

The role of the key multilateral institutions as 'handmaidens' to the interests of the First World thus looks set to continue in the future. The point is not that these institutions do not have power – on the contrary, this chapter has shown the knowledge and financial bases of the multilateral institution's power capabilities. Rather, it is that such power derives largely from the fact that these institutions 'represent' the interests and ways of thinking of the major First World states and, more generally, of the global capitalist system that those states are keen to promote. It is an exploration of the social and environmental impact of that system (and the role of businesses within it) in a Third World context to which this book now turns.

5

BUSINESS

The role of business in the Third World's environmental crisis has been a crucial one. Yet insufficient attention has been devoted in Third World political ecology to the analysis of this important type of actor. Instead, political ecologists have tended to focus on other actors, especially the state, in their work. A diversity of economic activities such as logging, mining or cattle ranching that contribute to environmental degradation have often been described with inadequate reference being made to the transnational corporations (TNCs) or local businesses that are often involved in those activities. Yet TNCs and local businesses have played a central role in the development of a politicised environment in the Third World, and appear set to increase that role even further in the near future.

This chapter seeks to explore the role of business in Third World environmental change and conflict. The growing power and influence of business today is linked to the development of a global capitalist system, a theme taken up in the first section of the chapter. The role of business within that system has been largely one of promoting economic practices that contribute to environmental degradation and social inequality in the Third World. First World-based TNCs have been the prototype in this regard, and the sheer size and economic power of these corporations helps to explain why attention has been focused on their perceived malpractices. Responding to their critics, First World TNCs have sought to relate their economic activities to sustainable development as part of the promotion of their 'green' corporate credentials, a theme also evaluated in this chapter. However, the political and economic prominence of First World TNCs in the Third World should not obscure (as is often the case) the important role of Third World businesses (hereafter referred to as local businesses) in the Third World's politicised environment. These corporations potentially enjoy selected advantages over their First World counterparts in the exploitation of the environment. These advantages merit consideration in any effort to understand the contribution of business to Third World environmental change and conflict. Further, a growing number of successful local businesses are branching out their operations into other countries, suggesting the emergence of a new phenomenon – the Third World-based TNC – in the Third World. The question as to whether the growing power of business in general represents a positive or negative influence on the quest for 'sustainable development' by many Third World countries concludes the chapter.

A GLOBAL CAPITALIST SYSTEM

A recurring theme in this book, and in the work of political ecologists generally, has been the role of a globalising capitalist system in the development of the Third World's environmental crisis. Chapter 3 noted the role of the state as a facilitator of that system in colonial and postcolonial times, while Chapter 7 considers the impact of capitalism on diverse grassroots actors, notably through a process of 'enclosure', whereby land and other key environmental resources are appropriated by states and large capitalist enterprises for intensive commercial exploitation. Here, however, the objective is to understand in general terms the dynamic of the global capitalist system, and the implications of that system for the role of business in Third World environmental change and conflict.

Many scholars have sought to explain the origins and expansion of the global capitalist system, and the role allocated within that system to the Third World (e.g Frank, 1967; Wallerstein, 1974; Wolf, 1982; Peet, 1991; Blaut, 1993). A general picture to emerge from such work is the uneven spread and impact of capitalism, as this 'mode of production' enters into contact with preexisting social and economic orders, and, along the way, transforms – but also adapts to – location-specific political, economic, social and ecological conditions. As Wolf (1982: 266) notes, 'driven by a general dynamic, capitalism yet gave rise to a variability of its own'. However, to speak of a global capitalist system is nonetheless to suggest a process whereby different location-specific conditions 'add up' to a 'whole' characterised by selected key themes (Peet, 1991).

One such theme has been the reorganisation of various Third World economies and societies in colonial and postcolonial times so as to facilitate production for a global market. As political ecologists point out, the role of the Third World in that market has been far from haphazard (Franke and Chasin, 1980; Watts, 1983a; Blaikie, 1985; Hecht and Cockburn, 1989; Rush, 1991). Rather, the tendency has been for Third World peoples and environments to be incorporated piecemeal into a globalising capitalist market in keeping with the needs and interests of capitalist enterprises based in Europe, North America and Japan. Until very recently, the role of the Third World was essentially confined to that of providing various natural resources to the First World, where those resources were then used to promote the development of industrialised societies. The Third World became the 'storehouse' of the First World providing such goods as cotton and groundnuts (Nigeria), copra (Philippines), teak (Burma), sugar (Brazil), rubber (Malaysia), tea and coffee (India) and copper (Zimbabwe). Indeed, such resource exploitation literally put the Third World on to the (colonial) world map, and in the process created local or national identities that are only now beginning to break down in those areas touched by the uneven spread of Third World industrialisation (Bryant and Parnwell, 1996). However, in those areas as yet untouched by industrialisation – which even today is most of the Third World – the resource identification still holds true.

A second theme pertaining to the global capitalist system relates to the central concern of businesses operating in the system to base their activities and market operations largely, if not entirely, on the logic of capital accumulation. Unlike many alternative forms of economic organisation (or modes of production), the drive to acquire capital has prompted the transformation of relations between people on the one hand, and between people and the environment on the other hand (Wolf, 1982; Peet, 1991). Thus, owners of capital (capitalists) have sought to control environmental resources, equipment and the labour of others so as to be in a position to capture the profit or surplus generated through production for the market. The power and wealth of capitalists derive precisely from an ability to control the 'means of production', and to deny to others access to those means. Thus, 'the way in which the [capitalist] mode commits social labor to the transformation of nature also governs the way the resources used and obtained are distributed among producers and non-producers' (Wolf, 1982: 77–8).

A third, and related theme concerning the global capitalist system is that the logic of capitalist accumulation leads directly to social and ecological 'contradictions' that may threaten the very fabric of life on earth as we know it (O'Connor, 1989). On the one hand, there is 'an appropriation of the earth's fecundity as a "natural resource" in the service of accumulation, and a running down of this resource' (O'Connor, 1994b: 5). This ecological contradiction suggests that environmental conservation ('sustainable devel-opment') and the global capitalist system are incompatible (Redclift, 1987). On the other hand, there is 'an appropriation of human nature, a domination exercised worldwide by capital . . . over humans qua labor and reproductive power' (O'Connor, 1994b: 5). This social contradiction suggests a basic antipathy between social justice and the global capitalist system. In aggregate, therefore, the quest to accumulate capital is not merely the appropriation of a surplus, but rather 'a destructive process whose result is the dereliction of human societies and ecosystems alike' (O'Connor, 1994b: 5).

A final theme relating to the global capitalist system is that this system is predicated on the elimination of most 'traditional' local environmental management practices by grassroots actors throughout the Third World as part of the integration of peoples and environments into a larger system over which they have no control (*Ecologist*, 1993). The flip-side of the control of the means of production by capitalists is the loss of control by grassroots actors over such means, and thus the loss of an ability to maintain an adequate livelihood independent of powerful outside actors. This theme is examined in detail in Chapter 7 in the context of a discussion of the connection between access, livelihoods and enclosure.

The preceding discussion has important implications for business, since this type of actor is at the heart of the global capitalist system. To begin with, the need for businesses to operate within this system tends, in practice, to favour those companies that have the requisite skills and networks to accumulate capital rapidly over those who do not. This point may seem very obvious, but it becomes important when considering why it is that business in most parts of the Third World has long been dominated by First World-based TNCs. The role of the latter in the Third World's growing environmental crisis is considered below, but here it is important to note that part of the explanation for the prominence of First World TNCs in the Third World today is simply a reflection of the 'headstart' enjoyed by those corporations as a result of their location in the First World, the 'home' of industrial capitalism. Concurrently, the tendency of European colonial rule in many parts of the Third World to eliminate indigenous business emphasises that there was nothing 'natural' about this process (Blaut, 1993). The end result has been a situation in which increasingly large and powerful First World firms have been able to conduct business in the Third World, sometimes virtually at will, but until very recently (and yet on a limited scale) their Third World counterparts have been too small to be able to do the reverse.

A further implication for businesses operating within a global capitalist system is that the system encourages them to acquire natural resources and labour as cheaply as possible in the quest to minimise production costs. Broadly speaking, the implications of this process for the Third World have been twofold. First, the quest for cheap natural resources almost inevitably led First World businesses to shift their focus from the First World, where natural resources were often already largely depleted by the early nineteenth century, to the Third World, where the supplies of many key resources used in industry have been in relatively plentiful supply (at least until very recently). The search by the British for timber in South and South-East Asia in the nineteenth and early twentieth centuries is a notable case in point (Grove, 1990). Second, the quest to obtain cheap labour has been one factor (among others) in the growing shift of manufacturing TNCs from the First World to the Third World since the 1960s. The argument that this move is also in part linked to environmental reasons – the so-called 'pollution haven' thesis – is explored below. The point here is simply that in the drive to minimise production costs, manufacturers in labour-intensive industries may have a strong incentive to shift part, or even all, of their operations to the Third World.

As part of the drive to reduce costs, businesses typically acquire a certain 'footloose' character in so far as they need to scour countries, if not entire regions of the world, for propitious production conditions. Those conditions vary both spatially and temporally, with the result that many medium to large businesses possess an itinerant status in which permanent allegiance is rarely given to any one locality. In this regard, such businesses (but not small local businesses) may be seen as the antithesis of the grassroots actors described in Chapter 7. Whereas the latter derive their purpose and power from their place-based qualities, the former gain their role in society from non-place-based qualities linked to capital accumulation. The footloose nature of businesses is most readily apparent in the case of TNCs (Korten, 1995, and see below), but may also apply within countries in the case of medium to large firms which shift operations and/or sources of supply in keeping with cost considerations (Gadgil and Guha, 1992).

Finally, the role of businesses within a global capitalist system is such as to predispose these businesses to cooperate with certain actors and to come into conflict with other actors. Thus, as Chapter 3 notes, businesses have often struck up close relationships with First and Third World states in the pursuit of their accumulation strategies. The role of the state as the provider of 'public goods' has been vital, as have its legal–coercive abilities vis-à-vis disaffected actors in society. As noted, the state–business alliance has not been without its tensions, as these two actors have similar, but not identical, interests. However, it is seemingly in the nature of a capitalist business to require certain 'goods' that only the state can provide, which serves as a powerful force for mutual accommodation between these actors (Johnston, 1989). Conversely, the interest of businesses in obtaining environmental resources and labour as cheaply as possible tends to embroil these actors in conflict with various grassroots actors who, from the point of business, are often obstacles to the 'free trade' in resources and labour. As Chapters 6 and 7 show further, from the vantage-point of most grassroots actors and environmental non-governmental organisations (ENGOs), businesses are potential, if not actual opponents (although growing links between some ENGOs and businesses suggest that this is not always the case).

The preceding discussion has sought to suggest briefly, and at a general level, the significance of the global capitalist system to the Third World's politicised environment, and the possible ways in which that system conditions the role of businesses as they pursue economic activities in relation to that environment. The remainder of the chapter examines more specifically the role of TNCs and local businesses in Third World environmental change and conflict.

TRANSNATIONAL CORPORATIONS

First World transnational corporations (TNCs) have been the principal driving force behind the globalisation of capitalism, especially over the last fifty years, and thus to consider the relationship between business and the Third World's politicised environment is to appreciate the impact of these corporations on human–environmental interaction in the Third World.

The sheer size and economic power of these TNCs is staggering. As the *Ecologist* (1993: 79) notes, 'seventy per cent of world trade is now controlled by just 500 corporations, which also control 80 per cent of foreign investment and 30 per cent of world GDP'. Much of that trade and investment occurs within the First World itself (Gill and Law, 1988). The role of TNCs is nonetheless considerable in the Third World, and the impact of TNC investments in the economies of Third World countries – typically smaller and more susceptible to TNC influence than First World economies – has been quite substantial. The contrast between these global economic Leviathans and most Third World states is striking. Thus, the gross income of Shell was $132 billion in 1990 – a figure that represented more than the combined gross national products of Tanzania, Ethiopia, Nepal, Bangladesh, Zaire, Uganda, Nigeria, Kenya and Pakistan in that year (countries with a total population of 500 million). Similarly, the gross income of the less well-known Canadian TNC, Cargill, equalled that of Pakistan (with a population of 100 million) in 1990, and this company controlled 60 per cent of the world cereal trade in that year (*Ecologist*, 1993). Faced with such discrepancies in economic power, it is hardly surprising that many Third World states have been in a weak bargaining position vis-à-vis First World TNCs (see below).

Yet the prominent role of these firms in much of the Third World since 1945 reflects the perception of both Third World states and First World TNCs that it is in their mutual interest to cooperate. On the one hand, the technological and financial power of these TNCs fits well with the development imperative motivating most Third World states over the last fifty years (see Chapter 3). The perception that TNCs would substantially enhance the development prospects of the Third World was a central tenet of the capitalist development theories that influenced Third World leaders in the 1950s, 1960s and 1970s (Corbridge, 1986; Peet, 1991). On the other hand, access to the Third World's cheap labour, emerging markets and environmental resources has been an integral part of the expansion plans of many First World TNCs. The relative weight attached to these factors in decisions by TNCs to establish operations in the Third World varies from company to company and from industry to industry, reflecting the complexities surrounding the decision-making process of large corporations (Pearson, 1987; Welford, 1996). However, a common theme has been the quest to obtain cost savings by moving operations to production sites in the Third World characterised by lax labour and environmental regulations (Korten, 1995). In this regard, the quest by TNCs to avoid increasingly stringent pollution control measures in the First World (linked in part to campaigns by First World ENGOs in that part of the world, see Chapter 6), has led to suggestions that they have chosen to relocate in the Third World in part because of their less stringent environmental regulations. Evidence for this 'pollution haven' thesis is widespread in the Third World (Leonard, 1988; Hardoy *et al.*, 1992).

A good example of this thesis is the case of the US-owned *maquiladores* (factories) located in northern Mexico along the border with the USA (see also Chapter 4). Since the early 1980s, some 2,000 American firms have set up operations here, contributing to local (if highly unequal) economic development and severe environmental degradation. Although these firms are required under Mexican law to 'ship their toxic waste back across the border to the USA...the heavily polluted local rivers and contaminated water supplies are powerful evidence that many are not doing so' (Coote, 1995: 31). Many of these businesses also emit highly toxic gases unchecked but, after all, the reason that many firms moved out of Los Angeles was 'to escape new restrictions on emission standards covering toxic solvents used in paints, stains, and lacquers' (Coote, 1995: 31). However, most TNCs move to the Third World for a variety of reasons, of which lax environmental regulations may not necessarily be the most important (Pearson, 1987).

Whatever their motives in establishing operations in the Third World, the contribution of First World TNCs to the Third World's social and environmental problems has often been immense. That contribution has been as varied as the types of economic activities that these businesses conduct in the Third World. Reflecting the traditional role of the Third World as a 'resource hinterland' of the First World (see above), the destructive impact of TNCs has been most noticeable in the natural resource-extraction industries – a theme prominent in the work of a number of political ecologists. Two examples suffice to illustrate the point. In Papua New Guinea, giant mining and logging TNCs are involved in the destruction of that country's tropical forests (Hurst, 1990). The giant Japanese Honshu Paper Company, for example, controls a vast 50,000 hectare timber concession in the Madang area, and has been responsible for much deforestation in that part of the country as a result of its unsustainable logging practices. Similarly, the American Amaco corporation acting in conjunction with the Papua New Guinea state is operating the OK Tedi gold and copper mine, one of the largest such operations in the world, but which has almost inevitably entailed widespread forest clearance in the area. The environmental effects of these operations have been many and have included loss of biodiversity, river siltation and poisoning (from the chemicals used in mining operations), soil compaction and depletion. The costs associated with such environmental degradation have been borne largely by local ethnic minority peoples dispossessed of their lands or unable to pursue traditional livelihoods due to the destruction of terrestrial and aquatic habitats (Hurst, 1990).

A second example relates to petroleum exploration and production in northern Ecuador by the US-based Texaco corporation acting in conjunction with Petroecuador, the state-owned petroleum company (Friends of the Earth, 1994; Kimerling, 1996). Since the discovery of oil in this remote part of the country in the late 1960s, oil production has increased, but at a growing cost in terms of local environmental degradation and social immiseration. Due to badly constructed and poorly maintained well sites, production pits and pipelines, oil spills and discharges of heavy metals and hydrocarbons have been common, with the result that local terrestrial and aquatic environments have been degraded. Since Texaco never bothered to develop a spill contingency plan, 'spills were not cleaned up' (Kimerling, 1996: 65). Infrastructure developments linked to oil production have resulted in localised deforestation, and access roads have enabled poor migrants to enter this hitherto inaccessible area, thereby compounding the latter problem as these settlers clear forest for agriculture. As in Papua New Guinea, the costs associated with such TNC-led 'development' have been borne largely by local indigenous peoples, primarily the Cofan, Secoya and Siona. Thus, as a leading actor in Ecuador's petroleum development programme, Texaco has left a legacy to the local people that 'includes environmental degradation, social disruption and cultural disintegration' (Friends of the Earth, 1994: 5).

These and other examples illustrate how, in the quest to accumulate capital, many First World TNCs have ignored environmental regulations (often with the active complicity of Third World states, see below), and in the process have contributed to social and environmental degradation in the production areas. Similar processes have occurred in urban areas due to the operations of manufacturing TNCs. The best known example here relates to the explosion of the US-owned Union Carbide chemical plant in Bhopal in 1984, noted in Chapter 2, in which at least 3,000 people were killed and some 100,000 were injured (Weir, 1988; Morehouse, 1994). The impact of the release of a toxic cloud of methyl iso-cynate into the air at the factory was not socially indiscriminate since it affected mainly low-income families living near the factory. Despite such human misery, Union Carbide has been able to parry all charges of corporate negligence in the courts with the result that those most affected by the industrial incident have received little or no compensation for their ordeal (Morehouse, 1994).

The impact of TNCs on Third World peoples and environments has been fairly evident. Less clear, however, is the degree to which these actors can be singled out unequivocally as blameworthy for the ensuing social and environmental degradation. As Redclift (1987: 73) observes, the argument that TNCs put 'profits before nature conservation and, for that matter, human livelihoods' is well established in the literature, but obscures as much as it reveals about their role and practices in the Third World context. Rather, it is important to examine the specific ways in which TNCs operate, and how their operations relate to the policies and practices of states whose responsibility it is to try and regulate them.

TNCs are the quintessential example of the 'successful' capitalist business enterprise, but the fact that their operations transcend national boundaries has important implications for the ways in which those operations impinge on environmental management and change in any given country. As Gill and Law note:

> In practice the transnational corporation, like other firms, is most concerned about making profits, corporate growth and increasing its market share. In pursuit of these goals its interests may diverge from those of both the host and the parent countries.
>
> (Gill and Law, 1988: 195)

Because they are disembedded 'from any one culture and any one environment, they owe no loyalty to any community, any government or any people anywhere in the world' (*Ecologist*, 1993: 79). TNCs seek to take advantage of the fact that they are non-place-based actors by pursuing their economic interests so as to maximise the potential gain to themselves, even if, as is often the case, this is at the expense of many place-based actors (Moody, 1996). A case in point is the routine practice of transfer pricing as a means by which TNCs maximise their overall profits, but to the detriment of most countries within which they operate. Transfer pricing occurs when products traded between branches of the same TNC but located in different countries are priced at a level which does not reflect the real market price of the goods. In countries where corporate taxation is high, prices are adjusted so as to minimise profits, or even show a loss in that country, thereby reducing, if not eliminating altogether, the tax bill that they must pay to the host state. To the extent that profits are registered at all, they tend to be concentrated in countries where corporate taxes are low (Gill and Law, 1988). In this manner, TNCs deny to Third World states revenue that is desperately needed for social development or environmental conservation projects.

Further, TNCs may hinder local business development in the countries in which they operate, undermining one of the main reasons for which Third World states often sought their investment in the first place. Taking advantage of their superior size, global connections and technical expertise, TNCs often seek to control each stage of the production process so as to maximise profits. This process excludes local businesses to the benefit of TNCs whose local dominance is thereby ensured. Taking the example of the Canadian grain company, Cargill, Kneen (1995: 195) notes that 'acting as input supplier, banker, buyer of finished products and whole-saler allows the company to make profits at every stage in the production, distribution and consumption of the commodities in which it trades'. Enhanced local dependency is the usual outcome. Thus, Cargill's 'export of cheap US grain to Third World countries has driven thousands of farmers off their land, unable to compete with the heavily-subsidized imports' (Kneen, 1995: 196). The environmental impact of this process is far from straightforward. Yet, to the extent that TNCs succeed in eliminating small businesses or grassroots actors from the production process, they may also be eliminating local management practices consonant with environmental conservation.

Indeed, the tendency of TNCs to adopt capital-intensive technologies in their economic activities may mean that the environmental management practices of these corporations are intrinsically more environmentally destructive than the practices of their local counterparts. While the use of capital-intensive technology may allow TNCs to maximise operational 'efficiency', such technology can be quite 'energy- and synthetic-intensive, and thus pollution-intensive' in comparison with alternative labour-intensive technologies (Gladwin, 1987: 6). The former has been associated with some of the worst cases of environmentally destructive economic activity in the Third World. In the forestry industry, for example, TNCs have been at the forefront in introducing capital-intensive mechanical logging in the Third World's tropical forests (Hurst, 1990). However, not only has this process undermined existing (often less destructive) forms of forestry, it has also set in motion a process of large-scale forest clearance and associated environmental destruction, which threaten to eliminate those forests in their entirety. Indeed, the large amounts of capital typically invested in 'modern forestry', and the fact that mechanical equipment needs frequent replacement, ensures that TNCs seek a quick return on their investments (Brookfield, 1988). The capital-intensive technologies that TNCs frequently use may not only be intrinsically more destructive than labour-intensive alternatives; they may also enable the perpetuation of destructive practices on a scale hitherto unknown in the Third World.

It could be argued that it is the task of individual Third World states to ensure that environmental regulations are in place to control potentially destructive TNC activities. Most Third World states certainly have a range of environmental regulations designed to specify appropriate business conduct in both the manufacturing and natural resource sectors (Gladwin, 1987; Brown and Daniel, 1991; Dwivedi and Vajpeyi, 1995). Many TNCs nonetheless continue to use practices which result in environmental degradation. To some extent, such continuity reflects the weak environmental management capabilities of many Third World states. As Gladwin (1987: 9) notes, 'it is important to recognise... that most developing countries face severe problems in the planning and management of natural systems and environmental quality and remain poorly equipped to protect and manage their environments'. To take again the Bhopal example noted above, the explosion at the Union Carbide plant occurred despite the existence of legislation on industrial safety, and notwithstanding provision in that legislation for annual inspections by state officials (Gladwin, 1987). For reasons noted in Chapter 3, states in the Third World have not accorded a high priority to environmental matters, and thus are often ill-equipped to regulate TNC operations (see also below).

TNCs often exacerbate this problem through efforts designed to frustrate all attempts by states to regulate their operations. Thus, TNCs may 'conspire to make purely national control systems either evadable, inefficient, incomplete, unenforceable, exploitable or negotiable (at the expense of desired environmental quality or occupational safety)' (Gladwin, 1987: 10). The rationale for such behaviour is clear enough. As noted above, the logic of capitalist business is to minimise costs so as to accumulate capital and, as we have suggested, First World TNCs moved to the Third World in the first place partly to escape stringent First World environmental regulations. The reasons why TNCs behave in this manner relate to the distinctive nature and power of these corporations.

TNCs are often able to use their footloose nature as a means to dissuade Third World states from enacting or enforcing strict environmental regulations (Korten, 1995). As Gladwin (1987: 10) notes: 'Faced with the prospect of onerous environmental regulations, a multinational can credibly threaten to close down a given plant, or to shelve expansion or investment plans there, and choose another national market for any additions to output.'

The threat by TNCs of a 'capital strike' gains added credibility in a context of pervasive Third World indebtedness and enforced structural adjustment policies. As Chapter 4 notes, the latter were a response by First World states, acting through the auspices of the World Bank and the IMF, to the Third World debt crisis. These policies have been predicated on neo-classical 'free market' theories that extol the virtues of global market competition and 'free trade' (Toye, 1993). Structural adjustment policies have been a boon to TNCs in so far as these policies have reduced the ability of most Third World states to control TNC operations, at the same time as they have placed a premium on inward investment – of which TNCs are the largest potential provider (George, 1992; Korten, 1995; Moody, 1996). Concurrently, these policies have given Third World states an added incentive to retain that TNC investment which is already in the country.

The financial plight of many Third World states enables TNCs to translate their immense economic power into political power through close political contacts at the senior levels of even the most powerful of Third World states. As noted below, there is often little need for TNCs to throw their weight around with Third World states due to a convergence of interests on environment and development issues. However, that convergence of interest ought not to obscure the ability of many TNCs to shape 'policy through their control of markets and the economic havoc they can cause by withdrawing support from a government ... [these] are the chief obstacles to the resolution of environmental and social problems' (*Ecologist*, 1993: 79–80).

Even if a Third World state chooses to enforce environmental regulations against TNCs, it faces formidable obstacles in ensuring that TNCs conform with the letter of the law. As a result of their footloose or itinerant status, there are definite 'limits on transnational corporate accountability and responsibility' (Gladwin, 1987: 12). As the example of Bhopal all too clearly shows, it is exceptionally difficult to make parent companies liable for destruction wrought by their subsidiaries in the Third World. In the Bhopal case, Union Carbide was able to contain legal proceedings in India, thereby avoiding potentially catastrophic litigation in the United States (Morehouse, 1994). The point here is not that TNCs are immune to regulation by Third World states, but rather that the latter face potentially important hurdles in efforts to do so.

However, as Chapter 3 notes, many Third World states have opted to woo TNCs with lax environmental regulations and other benefits as part of the quest for rapid economic development, thereby raising the issue of state complicity in TNC environmental malpractices. To a certain extent, such complicity has taken the form of official corruption in so far as state officials have accepted bribes in return for 'looking the other way' over TNC breaches of national environmental legislation (Rush, 1991). Yet state complicity runs more deeply than this in that it also encompasses a general reluctance to accord a high priority to environmental conservation measures – in other words, to echo an argument made in Chapter 3, states have been destroyers, and not stewards, of the environment. This point is nicely illustrated in an Indonesian context by Dauvergne (1993/4: 507) who notes that 'Indonesian central government decision makers view forests as a valuable, yet expendable resource, useful for generating foreign exchange to finance industrialization'. Thus, the alliance between states and TNCs in much of the Third World over exploitation of the environment rests on a combination of a quest for profits by TNCs and a desire for development by Third World leaders. It is this combined set of interests, and not merely those of TNCs alone, that is the source of much of the environmental degradation that plagues the Third World today.

State complicity is reflected in a general unwillingness by Third World states to adequately fund agencies empowered to enforce official environmental regulations. As Chapter 3 notes, environmental agencies are typically poorly staffed and thus ill-equipped to monitor potentially environmentally destructive activities by TNCs (and local businesses). To be sure, the recent surge of interest in 'sustainable development' has been reflected in the creation of a series of new environmental agencies which, on paper at least, possess considerable powers (MacAndrews, 1994). However, these agencies have often been adversely affected by structural adjustment inspired cuts in public expenditure and, even if spared such cuts, they have faced an uphill battle in persuading other agencies, let alone TNCs, to take environmental conservation seriously (MacAndrews, 1994; Dore, 1996; and see Chapter 3). The ability of TNCs to ignore environmental regulations derives in part from the political weakness of the environmental lobby within the state (although ENGOs may be changing this situation, see Chapter 6). It is in this context that Gladwin (1987: 234) has argued that 'the public policy environment must bear partial responsibility' for the Bhopal tragedy in India in 1984 in that understaffed and underfunded health and environmental safety agencies had been unable to conduct the regular safety inspections, required by law, that might have averted that disaster.

State complicity in TNC environmental malpractices also reflects a policy on the part of most Third World states to encourage TNCs (and local businesses) to invest in lucrative economic activities that simultaneously degrade the environment. States do so notably through a system of financial incentives (Repetto and Gillis, 1988; Hurst, 1990; Binswanger, 1991; Gadgil and Guha, 1992). As Repetto (1988: 2) notes in the context of tropical deforestation, these incentives have been 'sufficiently strong that forest resources have been destroyed for purposes that are intrinsically uneconomic'. The ways in which Third World states have sustained such potentially destructive activities as logging, mining, cattle ranching and intensive cash-crop production were considered in Chapter 3. What needs to be noted here is that TNCs have often (but not always, see below) played an important part in the expansion of those activities. They have therefore been important beneficiaries of state 'largesse'. In Indonesia, for instance, the advent of the New Order government headed by President Suharto in the mid-1960s was associated with a push to exploit the country's hitherto 'under-utilised' tropical forests. Large TNCs such as Weyerhaeueser (USA), Georgia Pacific (USA) and C. Itoh (Japan) operated large timber concessions in the forested Outer Islands (mainly Kalimantan) during the late 1960s and 1970s spurred on by a range of official incentives, especially five- and six-year 'tax holidays' (Gillis, 1988; Hurst, 1990). The activities of these firms enabled Indonesia to become a world leader in tropical timber exports – with a total log harvest in 1979 of 25 million cubic metres representing ten times the average annual production between 1960 and 1965 (Gillis, 1988).

Active state complicity at its most extreme may even extend to a willingness to provide direct military cover to TNCs in return for a share in TNC profits. The case of Shell Oil's petroleum operations in the Ogoniland region of southern Nigeria is apposite here (Rowell, 1995). This TNC has been engaged in petroleum exploration and development in the country since 1958, and oil production by this corporation (900,000 barrels per day in the mid-1990s) dominates both the national economy and state revenues (see Chapter 3). Shell's operations have caused considerable environmental damage in Ogoniland with the result that Ogoni activists, led by Ken Saro-Wiwa, mounted anti-Shell protests in the early to mid-1990s. The brutal response of the Nigerian military government involved the murder or arrest of Ogoni activists (including the execution of Saro-Wiwa), and a general campaign of terror by the dreaded Mobile Police Force in this region of the country. That the assistance of the security services was requested by Shell officials operating in the region – and even that this TNC may have been involved in illegal payments to the military ('protection money') – illustrates the often close link between TNC operations, state-sanctioned coercion, social oppression and environmental degradation (Rowell, 1995).

Yet if the Ogoniland episode illustrates the darker side of TNCs and their quest for profit, it would be wrong to suggest that the impact of TNCs on Third World peoples and environments is invariably negative. Political ecologists tend to emphasise the destructive nature of TNCs in the Third World (e.g. Dinham and Hines, 1983; Redclift, 1987; *Ecologist*, 1993). There is nonetheless a need for a more nuanced appreciation of TNCs that takes into account, for example, the potentially positive dimensions to their Third World operations.

Thus, the high profile of most TNCs may in fact encourage them to err on the side of caution in their environmental practices for fear of attracting bad publicity injurious to business. To the extent that TNCs investing in Third World countries may 'be under greater scrutiny and be more vulnerable to adverse publicity and punitive action' than is the case with their local counterparts, TNCs may be less inclined than the latter to conduct environmentally destructive operations (Gladwin, 1987: 6). It may not necessarily be the case as well that TNCs are able to use their economic power to browbeat Third World states into waiving environmental regulations that constrain TNC profit-making. Those states must weigh up a range of political and economic factors in their policy-making, of which the position and likely response of TNCs is only one consideration. Indeed, as noted below, local businesses may be able to escape regulations imposed on TNCs by virtue of their superior contacts in the state or as a result of their 'local' status. It is not surprising therefore that 'a host government under public pressure to do something about industrial pollution is likely to clamp down first on foreign industry' (Leonard and Duerksen cited in Gladwin 1987: 6). Foreign status may thus prove to be as much of a liability as it is a source of strength for TNCs. These considerations may prompt TNCs to conduct 'cleaner' operations than local firms due to the unique position of TNCs in Third World countries.

Further, TNCs may use 'environmentally sensitive' technologies in their operations that smaller local businesses are unable to afford. This argument seems to contradict the earlier claim that TNCs locate in the Third World precisely in order to avoid the need to use such technologies. However, the former view is linked to the idea that, as part of the development of a global production strategy, TNCs may use standardised technologies so as to benefit from economies of scale in their design and production. In the mining sector, for example, Brown and Daniel (1991: 46) argue that TNCs have been forced to develop 'clean' technology in the First World due to mounting environmental activism and official regulation there, but that this technology has thereafter become 'embodied in the investment package which substantial mining companies bring to exploration and development in developing countries'. As Gladwin (1987: 7) notes, there may also be a welcome demonstration effect in the process in that 'the MNC [multi-national corporation] can play, and many believe should play, a demonstrative and leadership role, and be an agent for change'.

Political ecologists are right to be sceptical of claims as to TNC virtuosity in light of their overall record in the Third World, but it would be unwise nonetheless for them to dismiss these arguments out of hand. To appreciate this point, it is necessary to consider the changing contemporary role of TNCs in relation to the growing worldwide emphasis on green development (Adams, 1990).

It is important to reiterate in this regard that, although TNCs may be very powerful, they are not immune from challenge or control by other actors (Schmidheiny, 1992; Eden, 1994; Welford, 1996). As concern over environmental degradation has mounted around the world, it has become imperative for TNCs to be seen to be taking an active stance on environmental problems. The point here is not that these corporations have suddenly converted to environmentalism based on the merits of the arguments of environmentalists. Rather, the prospect of altered TNC practices derives largely from the fear that, unless they alter their practices themselves, the tide of public sentiment against them will force states in the First and Third Worlds to impose stiff new environmental regulations inimical to capital accumulation. A related fear is that a failure to be pro-active will only assist ENGOs in their consumer boycott campaigns which threaten to hit TNCs where it counts – that is, in their 'pocket books' (Wapner, 1996; and see Chapter 6).

As a result, TNCs have paid increasing attention to environmental management issues since the mid-1980s. As noted below, they have reviewed individual corporate policies and practices, while simultaneously seeking to develop a collective corporate response to national and global debates over the definition of sustainable development. Most TNCs today therefore have well-developed corporate environmental strategies, and employ staff specifically to address the environmental implications of TNC operations. Concurrently, these corporations have gone to great lengths to publicise the 'greening' of their businesses as part of a long-term campaign to develop green credentials with officials and private citizens (i.e. consumers) alike (Welford, 1996).

To some extent, these business initiatives reflect an attempt to respond to the growing uncertainty that all businesses face in increasingly 'risky' societies, ironically shaped in part by the practices of those businesses themselves (Beck, 1992). As Chapter 2 noted, the systemic dimensions of a politicised environment reflect the growing ubiquity of 'unseen dangers' and health risks (i.e. nuclear fallout, pesticide build-ups in the food chain) associated with industrial production. Yet this process can boomerang on businesses in that the ability of these actors to pass on anonymously the social and environmental costs of industrialisation to other actors (especially poor and marginalised groups) is reduced because their products are publicly identified as being inherently hazardous. As Beck explains,

> Toxins, risks inherent in the product, counteract the constructions of anonymity. Thus palming off risks on the consumer becomes economically risky for the businesses themselves. They must also expect people to boycott their products, even where they are legally protected from liability claims.
>
> (Beck, 1995: 140)

These concerns, which appear at first glance to be mainly 'First World concerns', are becoming increasingly relevant in a Third World context in the measure that there has been a trend towards a standardisation in environmental policy by states around the world in the wake of the Rio Earth Summit of 1992 (see Chapter 4). The argument here is not that policy standards are identical in the Third and First Worlds – far from it. Rather, it is that there appears to be a gradual process of policy convergence taking place as a result of

> the internationalisation of the environmental interest groups, the creation of bilateral and multilateral environmental protection cooperation agreements, the policy harmonisation programmes in various regional areas, the sustained growth in international communications and contact among environmental professionals, a growing awareness of global interdependencies of many environmental problems, and the increased attention of some home governments to the environmental behaviour of their multinational corporations abroad (especially after the Bhopal tragedy).
>
> (Gladwin, 1987: 21)

To the extent that such policy convergence is taking place, the ability of TNCs to evade environmental regulations in the First World by simply shifting operations to the Third World begins to disappear. Thus, whatever the global distribution of their operations, TNCs are realising the need to develop a 'green' corporate strategy and image in their practices worldwide.

This perception is not confined only to those TNCs operating in the 'riskiest' industries (i.e. the chemical or nuclear industries). Rather, it represents an across-the-board change in attitude amongst TNCs, whatever the sector. For example, the United Kingdom based mining giant, Rio Tinto Zinc (RTZ), has been at great pains recently to promote an environmentally sensitive image, particularly in light of the criticism by ENGOs of its operations in Madagascar and elsewhere (Moody, 1996). The company publishes a glossy magazine (*RTZ Review*) that is designed in part to illustrate how RTZ incorporates environmental concerns into its mining activities, and it has even produced a video extolling the virtues of the company's environmental record. As one article in the magazine reported: 'It is RTZ corporate policy to carry out an environmental impact report on the development, operation and rehabilitation of a mine anywhere in the world, whether or not it is required by the local laws' (White, 1994: 5).

Using the example of the Morro do Ouro mine in Brazil which RTZ has operated since 1987, White (1994) notes that such environmental procedures were carried out, even though not required under Brazilian law at that time. The point here is not whether TNC practices conform to this rhetoric (see below), but rather that these corporations bother to mount such an expensive environmental campaign in the first place.

TNCs have also sought to cooperate amongst themselves in the pursuit of a collective response to national and global debates on sustainable development. The ultimate objective is to develop a 'common line' on environmental matters, and to use the combined power of these actors to shape the sustainable development agenda in keeping with business interests. Eden (1994), for example, has noted how, following publication of the Brundtland Commission report *Our Common Future* in 1987, TNCs cooperated with each other through the International Chamber of Commerce to produce a 'Business Charter for Sustainable Development' in 1991. This specifically acknowledged the need for businesses to adopt a series of 'sustainable' practices. The document was careful, however, to specify also that sustainable development was predicated on sustaining economic growth through business activity, thereby affirming the need for continued economic development (Eden, 1994). It was at the 1992 Earth Summit that the utility of this collective approach became fully apparent (Middleton *et al.*, 1993). There, TNCs were able to defeat the efforts of ENGOs and grassroots activists who sought to persuade states to tighten the regulation of TNC operations. As one disgruntled activist commented:

> the biggest flaw in the official documents which were being negotiated for signing by World leaders in Rio was the absence of proposals for international regulation or control of big business and TNCs to ensure that they reduce or stop activities harmful to the natural environment, human health and development.
>
> (Shiva, 1992: 106)

Indeed, TNCs won an important victory at Rio. Not only did the assembled world leaders (acting mainly at the behest of the United States) acknowledge that TNCs ought to regulate themselves, but they also suggested that these businesses should adopt a higher political profile in the future by contributing actively to the resolution of environmental and development problems at the local, regional and global scales (Shiva, 1992; Middleton *et al.*, 1993; Chatterjee and Finger, 1994).

The Earth Summit marked a high point in the TNC campaign to counteract the bad environmental publicity that has dogged these corporations for years. They have also sought to cooperate with other key actors over environmental matters. Relations with states have been further strengthened, and the latter increasingly seek 'the views of business' on how best to pursue sustainable development. However, TNCs have also sought to cultivate links with 'moderate' ENGOs so as to neutralise the threat posed to business from that quarter. As Chapter 6 notes, a number of leading ENGOs today depend for financial support on TNCs – the latter, ironically, once having been the target of criticism from the voluntary organisations. This process forms part of a larger attempt to 'divide-and-conquer' the ENGO sector by winning over moderate ENGOs to the business case, while attacking radical ENGOs not susceptible to the TNC 'charm offensive' (Stauber and Rampton, 1995).

TNCs have thus acquired a new political stature in global policy-making, but the preceding discussion would tend to suggest that they will use that stature in contradictory ways. On the one hand, they will need to elaborate a 'business vision' of sustainable development that somehow combines continued economic growth with selective environmental conservation. Otherwise, they will be liable to accusations of hypocrisy, and even worse, may face the unsavoury prospect of renewed regulation and escalating consumer boycotts of their products. On the other hand, they will also seek to develop that vision in such a way as to minimise the changes to the status quo, a situation, after all, that has enabled them to achieve unparalleled rates of capital accumulation. The broad aim is therefore to advocate incremental reform so as to ward off the prospect of major political and economic changes in the future. It would seem that TNCs will continue to play a central, if deeply contested, role in the Third World's politicised environment in the process.

Indeed, the ability of TNCs to condition that politicised environment is set to increase in the future. The case of the biotechnology industry illustrates this point. Biotechnologies have been heavily promoted by agro-chemical TNCs in recent years as a 'sustainable' form of agriculture (Kloppenburg, 1988; Kumar, 1993; Kloppenburg and Burrows, 1996). This new form of agricultural technology is based on the manipulation of the molecular structure and genetic makeup of plants, and promises to increase crop yields, even in ecologically marginal areas, without requiring (in theory at least) the addition of copious quantities of toxic chemicals of the Green Revolution technologies (see Chapter 4). As a handful of TNCs have monopolised this industry, fears have grown that 'research priorities in biotechnology are likely to be determined by commercial prospects and the global strategies of transnational corporations rather than by what is desirable for the poor in developing countries' (Kumar, 1993: 175). Such TNC control may result in technological packages that lock in the intensive use of chemicals as part of a TNC strategy to maximise control over all aspects of agricultural production. Corporations such as Monsanto, DuPont and ICI are 'contracting research to develop seeds resistant to their proprietary herbicides, hence making them more dependent upon chemicals' (Kumar, 1993: 175). There is thus the prospect that TNC domination of new biotechnologies may lead to the sacrifice of potential environmental gains in favour of continued TNC profits.

The preceding discussion has served to highlight the considerable economic and political power of TNCs in relation to Third World environmental change and conflict. However, local businesses are also playing an increasingly important role in such change and conflict, a topic to which this chapter now turns.

THIRD WORLD (LOCAL) BUSINESSES

Political ecologists have devoted relatively little attention to understanding how local businesses affect the Third World's politicised environment. Perhaps reflecting the structural roots of the research field (see Chapter 1), they have tended to emphasise instead the impact of First World TNCs on Third World environmental change and conflict as part of a broader argument about the place of the Third World in a First World-dominated global capitalist system (e.g. Franke and Chasin, 1980; Dinham and Hines, 1983). To the extent that local businesses have figured at all in Third World political ecology, they have done so typically in the context of discussions about state-sponsored economic activities that degrade the environment (e.g. cattle ranching). Yet the emphasis in these accounts has been on the role of the state in environmental degradation, rather than on the contribution of local businesses to such degradation per se (Repetto and Gillis, 1988). This section suggests, in contrast, that the impact of local businesses on the Third World's politicised environment has been greater than hitherto portrayed. Indeed, these firms may play in aggregate an even more critical role in the evolution of that environment than do the TNCs.

Part of the reason why political ecologists tend not to focus on local businesses may be that the nature of these firms' activities is often harder to classify and interpret than is the case with TNCs. While the latter are typically large, well-defined organisations, the former range from small firms employing only a handful of employees to huge conglomerates which may rival First World TNCs in terms of staff and annual turnover (indeed, an increasing number of the largest such firms are becoming TNCs themselves: see below). To understand the role of local businesses in Third World environmental change and conflict is thus potentially a 'messy business' analytically. Nonetheless, and in light of our discussion of TNCs above, three general points about that role merit brief comment here. First, the size of local businesses varies greatly from country to country and from industry to industry. In certain sectors (e.g. electronics or pulp and paper), technical or financial barriers to entry limit the number of potential entrants to the industry, while in other sectors (e.g. rubber or surface gold mining) small businesses may flourish. Local business activity will also vary depending on the ideological disposition of the local state towards the business sector in general, and towards TNCs versus local businesses in particular.

Second, and notwithstanding the above, the trend in many Third World countries has been towards the domination of various sectors by larger and larger local businesses (perhaps acting in conjunction with First World TNCs, see below). Thus, while First World TNCs may be playing an increasingly prominent role in the Third World today, this trend has not necessarily been at the expense of the more successful local businesses. Indeed, in an era of structural adjustment and state retrenchment (see Chapters 3 and 4), local businesses and TNCs may both be benefiting from 'the privatization of everything' (Watts, 1994), as they increase their wealth and power vis-à-vis both local states and small local firms. This process of consolidation must certainly not be exaggerated since small businesses may possess selected political and economic advantages over their larger compatriots or even over TNCs. To take but one example, efforts by large national companies, as well as the United Kingdom-based RTZ corporation, to prevent small-scale mining operations in Brazil's Amazonian gold fields have come to nought over the years as *garimpeiros* (gold miners) and *donos* (small entrepreneurs) have lobbied successfully for their economic rights (Cleary, 1990). Nonetheless, industry consolidation does appear to be a general trend in most sectors of the economy as the quest to accumulate capital in a global economy intensifies.

Third, the 'local' nature of local businesses has meant that these firms have needed to place a premium on winning state support to an extent not usually required of TNCs. The point here is not, of course, that TNCs do not require such support. Rather, it is that local businesses cannot simply move to another country, and must as a consequence devote greater attention than TNCs to the cultivation of political contacts within the state as part of their accumulation strategies. Indeed, it is this nexus of state–business interests that has been most widely remarked upon by political ecologists (Schmink and Wood, 1987; Hecht and Cockburn, 1989; Hurst, 1990). To what extent the contemporary 'retreat of the state' and the globalisation of economic activity has lessened the need for local businesses to solicit state support so assiduously remains to be determined.

Notwithstanding the diverse ways in which local businesses relate to each other and to other actors, these firms have played an important role in the evolution of the Third World's politicised environment. A central issue in this respect is whether that role has been more or less destructive environmentally than that of TNCs. Rather than attempt a crude quantitative measure of the relative impact of TNCs and local businesses, this issue is used here as a way in which to consider the potentially distinctive political, economic and cultural factors that might influence the latter's role in Third World environmental change and conflict.

Several factors suggest that local businesses may have a less severe social and environmental impact than is the case with TNCs. To begin with, local businesses are typically smaller and more labour-intensive than their transnational counterparts. Gladwin (1987: 6) suggests that this trait is especially important in that, since 'the production processes of local enterprises tend to be less capital-intensive ... locals may rely less on production processes that are energy- and synthetic-intensive, and thus pollution-intensive'. This 'small-and-labour-intensive-is-beautiful' argument is considered further below, but the relative absence of capital-intensive means of production in both the manufacturing and natural-resource extracting industries may indeed limit the possible damage that any one local business is able to inflict on the environment. Of course this argument does not apply in the case of larger and more capital-intensive local businesses – pulp and paper corporations in southern Brazil being a case in point (Marchak, 1995).

Local businesses may also adopt relatively restrained environmental practices as a result of their local political, economic and cultural 'embeddedness', and the attendant fear of adverse public and state reactions to environmentally degrading practices. Unlike TNCs, local businesses are unable to leave the country in the event of adverse publicity or the threat of state sanctions, and must accept the local political and economic consequences of their actions. As we note below, such considerations have rarely weighed heavily in the strategic calculations of local businesses. Yet the spread of democratic or quasi-democratic political conditions, as well as the rise of ENGOs and grassroots organisations in many parts of the Third World since the 1980s in particular, suggests that the social and environmental consequences of local business activity is coming under ever greater scrutiny and criticism (see Chapters 6 and 7). Local businesses, as with TNCs, are increasingly facing up to the need to develop a 'green corporate strategy', but with the added impetus of the prospect of diverse and potentially inescapable local penalties (i.e. consumer boycotts, increased regulation) in the event that they fail to do so. As with TNCs, however, it is an open question as to whether the 'greening' of local businesses is anything other than rhetoric masking a business-as-usual approach (see below).

Indeed, diverse factors suggest that local businesses may be even more destructive environmentally than TNCs. First of all, the former may be better placed than the latter to avoid environmental regulations. On the one hand, TNCs are condemned by their political and economic status to being 'trapped in the public gaze' to an extent that is rarely, if ever, the case for local businesses. As a result, while local businesses may be able to evade costly regulations with the collusion of state officials without risking a major public outcry, the likelihood that TNCs would be able to pursue a similarly surreptitious strategy is increasingly remote. Further, local businesses do not also have to worry about the reaction of the public in the First World as is increasingly the case with TNCs. On the other hand, local businesses may have a close rapport with state leaders and officials as a result of their local embeddedness (encompassing even family connections) which TNC executives could not even hope to match. As a result of these close ties, local businesses, in contrast to First World TNCs, 'may enjoy greater influence with governmental regulators and may be granted preference with respect to the stringency and timing of environmental performance standards' (Gladwin, 1987: 6).

Second, local businesses may also be more adept than TNCs at capturing local state subsidies to undertake environmentally destructive activities. As political ecologists have shown, for example, cattle ranching in Brazil's Amazonia has been a large-scale private sector business activity that has been wholly predicated on cheap loans and other financial incentives provided by the Brazilian state (with some assistance from international financial institutions, see Chapter 4). As Schmink (1988: 167) notes, 'most business ventures in Amazonian forests are controlled not by foreign but by domestic investors'. Indeed, through personal ties to the state, and through lobby groups such as the powerful Association of Amazonian Entrepreneurs, domestic investors were able to extract from a staunchly pro-business military government more than one billion United States dollars in the space of only fifteen years in the late 1960s and 1970s for cattle ranching in the Amazon (Schmink, 1988). The state subsidised the move by many large Brazilian businesses to acquire land and diversify their investments, but in a process which underpinned an ecologically and economically unsustainable activity across a large part of the Amazon (Bunker, 1985; Hecht, 1985; Browder, 1988; Hecht and Cockburn, 1989).

Finally, the labour-intensive practices of local businesses can frequently be more environmentally destructive than the capital-intensive practices used by their transnational counterparts. Local businesses may simply not be able to afford the requisite technology that would enable them to undertake economic activities in an environmentally sensitive manner (Brown and Daniel, 1991). The case of small-scale informal sector Amazonian gold mining alluded to above is a case in point. Thus, Durning notes:

> In the late eighties ... tens of thousands of gold prospectors infiltrated the northern Brazilian haven of the Yanomami, the large, isolated group of indigenous peoples in the Americas. The miners turned streams into sewers, contaminated the environment with the 1,000 tons of toxic mercury they used to purify gold, and precipitated an epidemic that killed more than a thousand children and elders.
>
> (Durning, 1993: 86)

The vast majority of the *garimpeiros* and *donos* involved in this process were unable to afford the necessary equipment to treat adequately and safely the toxic mercury used in their activities, even when acquainted with the grave environmental and health risks associated with such activities (Cleary, 1990). Thus, 'small-and-labour-intensive' is not necessarily beautiful in relation to the environmental practices of local businesses.

Yet the ability to afford 'clean' technologies does not mean that businesses will necessarily use those technologies (Hurst, 1990; *Ecologist*, 1993). This point was made earlier with reference to TNCs, but here, and sticking with the example of mining, the role of larger and more capital-intensive local businesses is considered. Many of the latter have prospered in partnership with First World TNCs, and Benguet Corporation, one of the Philippines' most powerful corporations, is no exception to this rule. This corporation, which was founded by American investors in 1903 but is 'now owned in approximately equal thirds by wealthy Filipinos, the Philippine government, and US investors', operates the huge Grand Antamok open-pit gold mine in the Cordillera Mountain Range on the island of Luzon (Broad, 1993: 29). While this mine, and other mines worked in the Philippines by the corporation, have contributed substantially to the country's minerals export trade, Benguet's mining practices have resulted in severe environmental degradation including localised deforestation, depleted water tables, and the poisoning of local rivers as a result of the routine dumping of toxic chemicals (used by the company to separate the gold from the mined deposits) into water courses by company officials. Further, and mirroring comments made earlier with regard to TNCs, the costs associated with such degradation are borne disproportionately by those poor and marginalised actors who happen to live in the area in which the Grand Antamok mine is located. Thus, the environmental problems noted above have adversely affected the livelihoods of the local Igorot people, an ethnic minority group long resident in this area. For example, the ability of the Igorot to grow rice on the terraces constructed on the mountain sides by their ancestors has been greatly reduced due to poisoned or depleted water supplies. Further, the traditional small-scale 'pocket' mining of these people has been disrupted by the large-scale open-pit mine operated by the Benguet Corporation (Broad, 1993).

The 'local' status of businesses such as the Benguet Corporation may prove to be a bigger obstacle to the efforts of ENGOs and grassroots organisations to end such destructive business practices than is the 'global' status associated with TNCs. Not only do these local firms possess effective political contacts within the state (as the Benguet example illustrates, the state may even be a shareholder), but they are also able to argue that their activities are 'in the national interest' in a way that TNCs are simply unable to do. Concurrently, powerful local businesses can attack ENGOs and grassroots organisations as a 'disruptive influence' in society (in Cold War times, the Communist label was also typically used).

Local businesses may thus be even more destructive environmentally than many First World TNCs (although further research is needed on this topic). However, two trends in the development of local businesses will increasingly affect the ways in which these businesses conduct their activities, albeit in potentially contradictory ways.

One trend is the emergence of Third World-based TNCs as a force to be reckoned with at the global level, thereby raising the intriguing question as to the likely impact of these corporations in countries other than their own. It might be anticipated that Third World TNCs might seek to avoid the environmental mistakes of their First World counterparts by implementing sustainable practices in host countries so as to avoid public opprobrium. Although it is still early days, the available evidence would tend to belie that assumption. In the case of the forestry industry, for example, East and South-East Asian TNCs have acquired logging concessions in other parts of the Third World, but their practices in these areas have been condemned as socially and environmentally harmful. Colchester (1994b) thus reports that South Korea's Sung Kyong trading company and Malaysia's Samling Timbers corporation now dominate the expanding logging industry in Guyana, encouraged notably by a heavily indebted local state desperate to raise foreign revenue. However, these Asian corporations are repeating the unsustainable logging practices that they have previously used in their home countries in this politically isolated and weak Latin American country. In the quest to accumulate capital as rapidly as possible, therefore, it would appear that there are few grounds for expecting Third World TNCs to be any less destructive environmentally than First World TNCs in the Third World.

A countervailing trend, however, has been the growing need for businesses in the Third World to defend and justify the environmental impacts of their activities. Thus, despite practices such as those noted above, Third World TNCs have nonetheless begun to follow their First World counterparts in developing a green corporate image for themselves. To some extent, this strategy has mainly reflected the interests of those Third World manufacturing TNCs competing in environmentally conscious First World markets. Yet, as the next two chapters in particular show, environmental consciousness is not a monopoly of the First World. Indeed, a ubiquitous feature of domestic politics in most parts of the Third World today is the efforts of ENGOs and grassroots organisations to end the socially and environmentally destructive activities of businesses and states. In the measure that these efforts succeed, all Third World businesses will need to pay greater attention to the social and environmental implications of their actions. The result may even be the spread of sustainable practices among local businesses in the Third World.

PARTNERS OR VILLAINS IN SUSTAINABLE DEVELOPMENT?

This chapter has explored the role of TNCs and local businesses in the evolution of the Third World's politicised environment. Two themes in particular emerge from the discussion. First, both local and transnational businesses have played an important role in the shaping of that environment. Indeed, in 'an age of market triumphalism' (Peet and Watts, 1993) that role is set to become even more central. Second, local and transnational businesses have generally had an adverse environmental effect, because these firms privilege profit maximisation over social justice and environmental conservation in their day-to-day operations.

The poor environmental record of businesses sits uneasily with the idea that, post-Rio, these actors are now 'partners in sustainable development'. Seemingly akin to 'putting the foxes in charge of the chicken coop', this strategy reflects, as we have noted, the growing ability – especially of TNCs – to translate economic power into political power at local, regional and global scales. Yet it is not at all clear at the present juncture how businesses will use their enhanced political and economic power. Left to their own devices, they would undoubtedly seek to ignore environmental regulations since it remains in their immediate economic interest to do so. By continuing to pass environmental costs on to society at large, they minimise operation costs and thereby maximise corporate profits. As we noted at the start of this chapter, there seems to be an inherent contradiction between the quest to accumulate capital and the need to conserve the environment. Yet, however powerful they may become, businesses will never be completely free to do as they please. Businesses may be increasingly powerful actors in Third World environmental change and conflict, but as this book shows, they are not the only actors with power in these processes. Their ability to condition the evolution of the Third World's politicised environment may have grown, but that ability is far from complete.

Two further issues remain to be considered here concerning the role of business in shaping that environment. The first issue relates to the question of whether businesses can be persuaded that sustainable development is worth pursuing in its own right, on the basis that it might afford them glittering new opportunities for capital accumulation. In other words, can appeals to financial self-interest succeed where moral appeals fall on deaf ears? The evidence so far is decidedly ambiguous. A growing number of local and transnational businesses are pointing to their involvement in 'green' business activities such as plantation forestry or eco-tourism as evidence that they are 'leading the way' in the quest for sustainable development. At first glance, such activities would appear to be an ideal marriage between the principle of profit maximisation on the one hand, and environmental conservation and social justice on the other hand. Yet the public apparently needs to be primed as to the merits of these activities. Fernandez Carro and Wilson (cited in Lohmann, 1996: 23) speaking about plantation forestry note that 'success is measured by the freedom to plant fibre crops...Our objective should be to create and move inside an ever-increasing friendly circle of public opinion' (see also Stauber and Rampton, 1995). It would therefore appear that the definition of economic activities deemed to be sustainable by businesses does not square with the perceptions of the public at large. Indeed, as Chapter 3 noted, political ecologists have already pointed out the ways in which state-business efforts to promote green activities, such as eco-tourism and plantation forestry, have led to mounting protests by grassroots organisations and ENGOs about the perceived social and environmental inequities of these ostensibly green activities (e.g. Hirsch and Lohmann, 1989; Peluso, 1993b; Lohmann, 1996).

The second issue follows on from the first point, and is related to the prospect that other actors may be able to force recalcitrant local and transnational businesses to pursue environmentally sound practices through diverse political strategies. This claim may seem paradoxical given the argument of this chapter that the power of business has grown in recent years. However, and as discussed at greater length in the following two chapters, the 'empowerment' of civil society vis-à-vis the state has increased the power not only of businesses but also of ENGOs and grassroots organisations (Peet and Watts, 1993). In a sense, the growing assertiveness and power of the latter two types of actors is directly related to the widespread sentiment that neither the state nor business is capable of promoting sustainable development since both of these actors have benefited so handsomely from environmentally destructive activities in the past.

On the surface, the battle between ENGOs, grassroots organisations and states-businesses would appear to be a struggle between David and Goliath. Yet, as we noted in Chapter 2, power can be much more than simply a matter of totting up material resources – it also embraces the realm of ideas. It is this latter consideration that may enable ENGOs acting in conjunction with grassroots organisations to force businesses into altering their activities.

This chapter has shown how businesses operating in a global capitalist system premise their actions on the promotion of capital accumulation. Moral considerations of social justice and environmental conservation do not figure in these strategies per se. However, and as we shall see in the next chapter, the power of ENGOs derives precisely from moral considerations, and their ability to speak to social and environmental issues beyond the realm of the capitalist market. These actors are increasingly able to translate intangible moral considerations into tangible economic considerations, notably through the weapon of the consumer boycott campaign. In the measure that they succeed, ENGOs may be in a position to tap growing popular anxiety about the environment in both the Third and First Worlds so as to leave businesses confronted with the prospect of mass boycotts of their products with little choice but to 'change their spots'.

The point here is not to exaggerate the ability of ENGOs or grassroots organisations to change the ways in which either businesses or states operate. Rather, it is simply to reiterate a theme that appears again and again throughout this book – namely, that, while relations between actors is conditioned by unequal power relations, no actor is omnipotent and hence no actor is completely powerless. The implications of this theme in terms of the struggles of ENGOs and grassroots organisations on behalf of the livelihood concerns of poor grassroots actors are taken up in the next two chapters.

6

ENVIRONMENTAL NON-GOVERNMENTAL ORGANISATIONS

The growth of environmental problems worldwide has been associated with the emergence of a new type of actor – the environmental non-governmental organisation (ENGO) – in political-ecological conflicts. Although few parts of the world have been untouched by this process, it is arguably in the Third World that the political impact of ENGOs has been the greatest. Accordingly, the influence of this type of actor on the changing topography of the Third World's politicised environment is coming under increasing scrutiny by policy-makers and scholars alike.

This chapter explores the growing significance of ENGOs in terms of the Third World's environmental crisis. At a general level, the rise of ENGOs can be seen to reflect the growing power and assertiveness of 'civil society' vis-à-vis the state in most Third World countries since the late 1970s. Yet just as the term 'civil society' encompasses a diversity of social groups and interests, so too the term 'ENGO' also incorporates a wide range of social concerns and practices (Meyer, 1995). We would suggest that it is useful in attempting to understand the political-ecological significance of ENGOs in the Third World to focus on two broad types of ENGO: First World-based advocacy ENGOs (most of whom are animated by First World environmental concerns) and Third World-based ENGOs (many of whom are centrally concerned with basic livelihood issues). As discussion of the latter will highlight, 'environmental' NGOs in the Third World are mainly concerned with development issues, notably the promotion of social justice and equity for poor marginalised grassroots actors. Yet, what often distinguishes them from regular 'development' NGOs, such as Oxfam or the Grameen Bank, is their emphasis on the need to pursue such objectives via the mechanism of environmental conservation. Social justice and equity is attained by ensuring that the poor gain access to local environmental resources (i.e. timber, fuel, clean water). While development NGOs may address these concerns from time to time, their focus nonetheless lies elsewhere – viz., health care, education or famine relief. These NGOs thus lie beyond the remit of a chapter that is centrally preoccupied with explaining how ENGOs affect Third World environmental change and conflict.

Following a discussion of the role of ENGOs in civil society, our attention turns to the analysis of First World advocacy ENGOs and Third World ENGOs. The relationship between ENGOs and states is thereafter considered as part of an evaluation of the broader issue of whether or not the advent of ENGOs signifies a new politics of the environment in the Third World.

ENVIRONMENTAL NGOs IN CIVIL SOCIETY

Since the end of the 1970s, there has been a massive increase in the size and number of non-governmental organisations worldwide of which the ENGO is but one type. At a general level, this tremendous growth reflects the increasingly central role that civil society is taking in the management of various aspects of social and environmental well-being. That increasing role, in turn, is linked seemingly to the declining capacity of the state as an actor to provide for such well-being. This book has already shown why the state has failed for various reasons to live up to its self-proclaimed role as both developer and steward of the environment (see Chapters 3 and 5). Yet it has been a widespread public perception that states have contributed to – rather than mitigated – poverty and environmental degradation that has played an important part in encouraging diverse groups in civil society to become more assertive over social justice and human rights issues. This process has occurred unevenly and is linked in particular to the spread of democracy in many parts of the Third World since the 1980s.

A large and eclectic theoretical literature now exists on the inter-relationship of the 'new social movements', civil society 'empowerment' and democratisation (e.g. A. Scott, 1990; Clark, 1991; Friedmann, 1992; Peet and Watts, 1993, 1996b). For our purposes, it is useful to focus on the potential political significance of ENGOs, as well as on their great diversity in terms of organisation, purpose and social basis.

The sources of the political power of ENGOs elude easy description. Unlike the state, ENGOs do not possess a formal monopoly on the means of coercion within a defined territory and, unlike some businesses, they typically do not control sizeable amounts of capital. Lacking these political and economic advantages, ENGOs are nonetheless formidable players in environmental conflict and change in the Third World (Meyer, 1995). This is the case because ENGOs possess a strong 'moral' character seemingly absent in most other actors. As Princen (1994a: 34) suggests, this type of actor can be distinguished from others in that its activities reflect 'qualities of legitimacy, transparency, and transnationalism'. Environmental NGOs are seemingly unprepared to sacrifice environmental quality for the often ephemeral benefits associated with economic growth in the way that states, businesses and multilateral institutions, for example, frequently seem willing to do. Thus, 'relative to actors in the governmental and business sectors, in the environmental realm, NGOs are perceived as defenders of values that governments and corporations are all too willing to compromise' (Princen, 1994a: 34–5).That ENGOs are not seen to profit from activities that degrade the environment unlike some powerful actors – indeed, that their purpose is to seek ways in which to solve environmental problems – provides them with a reserve of public goodwill that they are able to use politically. The rest of this chapter illustrates this point with reference to First World and Third World ENGOs, but here the general political techniques used by these actors merit comment.

The first way in which ENGOs seek to exert political influence is through an attempt to influence the environmental policies and practices of states, businesses and multilateral institutions. As we shall see, the ability of ENGOs to lobby these powerful actors successfully, and to encourage major policy changes, has been quite considerable in recent years. A number of large international ENGOs have, for example, worked together to persuade First World states and multilateral institutions (e.g. the World Bank) to abandon economic aid packages to the Third World which are predicated on environmentally destructive development (Hurrell, 1992; Rich, 1994; see also Chapter 4).

A second way in which ENGOs exert political influence is through direct action linked notably to the support of conservation and development projects proposed and operated by grassroots actors. The link between grassroots actors and ENGOs is explored below and in Chapter 7, yet the point here is a general one – namely, that by bypassing the state and other actors in this manner, ENGOs are often making a highly political intervention that may have great implications for power relations at the local level (Peluso, 1995). In so far as such intervention results in the protection of a given environment or alleviates the plight of marginal actors, ENGOs challenge the interests of those actors who support the status quo. However, and as is also noted below, the interests of ENGOs and grassroots actors are not necessarily compatible (Thrupp, 1990; Peluso, 1993b).

A third way in which ENGOs seek political influence is through well-publicised campaigns (notably through the media) that are designed to raise public awareness on environmental issues. This process may not only lead to the growth of 'sustainable' consumer practices (e.g. recycling), but also helps ENGOs to sway the actions of traditionally powerful actors, especially states. A big part of the success of ENGOs is related to the ability to use the media and local networks to promote various campaigns. Large international ENGOs are particularly able to 'command media attention on some issues in ways that few other actors can' (Princen, 1994a: 34). However, even small ENGOs have well-developed regional and local networks through which they can disseminate their message effectively to local communities in a much more effective manner than other actors.

Finally, ENGOs seek to exert political influence through activities at global environment and development conferences. Chapters 3 and 4 note that these conferences are an important forum for inter-state conflict, negotiation and agreement on environmental matters. A further trait of the conferences is the presence of a large contingent of ENGOs (mainly from the First World), which serves to remind a watching world that alternative views on environment and development issues exist to those propounded by state or business leaders (Chatterjee and Finger, 1994). It is an ideal occasion for ENGOs to criticise the 'parochial' views of assembled state leaders, and to contrast those views with their own ideas on the world's environmental problems.

The potential political influence of ENGOs needs to be set against the fact that a great variety of groups operate under the 'ENGO' label (Smillie, 1995). Environmental NGOs are highly differentiated according to organisational size, structure, skills, purpose, ideology, cultural origin or legal status (Branes, 1991; Livernash, 1992; Clark, 1995). For example, there are a small number of very large transnational ENGOs, such as Friends of the Earth (FOE), Greenpeace or the World Wildlife Fund (WWF), which employ hundreds of employees, manage multi-million dollar budgets and maintain offices around the world (Wapner, 1996). In contrast, there are thousands of small ENGOs that are kept going by a handful of dedicated staff operating on a shoestring budget (Fisher, 1993). Many ENGOs also maintain close political and economic links to states and businesses – links that sit uneasily with the image of ENGOs as independent actors (Meyer, 1995; and see below).

Environmental NGOs are also differentiated according to whether they are membership support or non-membership support organisations (Farrington and Bebbington, 1993). Many First World international advocacy ENGOs discussed below are membership-support organisations with direct and permanent links to the public. Competition for public support amongst these ENGOs can be fierce since successful organisations gain considerable financial clout and political credibility from their membership. Greenpeace, for example, is one of the largest ENGOs in the world with over 6 million members and an income estimated at more than $100 million in 1994 (Wapner, 1996). In contrast, other ENGOs may be classified as non-membership support organisations that often derive financial resources directly from states or businesses. These ENGOs notably include 'think-tank' organisations which exert political influence through the publication of scientific findings (Bramble and Porter, 1992). The Washington-based World Resources Institute (WRI) is one such ENGO and publishes research regularly on the 'state of the world's environment' (e.g. World Resources Institute, 1990), as well as working on specific environmental issues (Repetto and Gillis, 1988).

Environmental NGOs differ according to the scale(s) at which they operate. Large ENGOs such as Conservation International or Greenpeace operate at all scales, addressing a wide range of issues (e.g. deforestation, nuclear testing, global warming) in the process. In contrast, many small ENGOs tackle highly specific environmental problems at the local scale only. For example, in the Philippines and India alone, several thousand organisations fit into this category (Rush, 1991; Broad, 1993; Fisher, 1993). A recent development in many Third World countries has been the creation of national ENGOs and even national coalitions of ENGOs to develop an indigenous nationwide response to environmental problems. In Ecuador, for example, the *Fundación Natura* (Nature Foundation) is a national ENGO that seeks to protect the country's rich ecological heritage through lobbying, public campaigns and projects throughout the country (Meyer, 1995; Bailey, 1996). In the Philippines, the Green Forum-Philippines is a national policy advocacy organisation representing the interests of member ENGOs in campaigns that target state agencies, businesses and multilateral institutions (Legazpi, 1994).

Environmental NGOs also vary according to their cultural origins and the socio-economic context in which they develop. The most striking difference is between First World and Third World ENGOs. The former are generally products of the 1960s protest movements in North America, Europe and the Antipodes, and reflect aesthetic values about the 'natural' environment and its preservation (McCormick, 1995; Wapner, 1996). Third World ENGOs, in contrast, often developed out of the livelihood concerns and interests of local communities threatened with social and environmental degradation arising from the actions of states or businesses (Redclift, 1987; Chatterjee and Finger, 1994). Different organisational traits are usually associated with these different development trajectories. As Middleton *et al.* (1993: 178–9) note, First World ENGOs have typically developed as a reflection 'of the confidence of their industrial worlds' in that they 'tend to be professional, to have well-organised methods of fundraising and a management approach to problem-solving'. These traits are reflected in how most First World ENGOs seek to 'deal with' the Third World's environmental crisis. They promote a 'professional approach', working directly with Third World states or ENGOs through the provision of technical, financial or human support. In contrast, many Third World ENGOs have developed in a less formal or 'professional' manner. Indeed, those organisations closest to grassroots actors ('grassroots organisations') often employ local community members who frequently lack the training and professional skills possessed by the staff of First World ENGOs (see Table 6.1 and Chapter 7).

However, not all Third World ENGOs fit this description. Thus, one of the more interesting recent developments in the Third World ENGO community has been the rise of indigenous advocacy organisations and grassroots support organisations (service organisations) in response to the environmental anxieties of the Third World's expanding middle classes (see Table 6.1). Although most Third World ENGOs are community-based grassroots organisations dedicated to the direct protection of local livelihoods, the number of advocacy or grassroots support organisations is growing rapidly, especially in the more economically prosperous Asian and Latin American countries (Redclift, 1987; Rush, 1991). These ENGOs, reflecting their middle-class origins, tend to emphasise environmental quality issues (e.g. air pollution, biodiversity), and are usually highly professional, staffed with trained experts and rarely representative of the communities in which they work (Middleton *et al.*, 1993). In many cases, they also work closely with their First World counterparts or with multilateral institutions, pursuing jointly agreed projects, and receiving considerable financial and technical support from these actors in return. Yet, as its name would suggest, the grassroots support organisation in particular seeks to integrate the environmental conservation concerns of its middle-class supporters with the basic livelihood issues of the (poor) communities within which this type of ENGO works (Carroll, 1992).

Two contrasting examples serve to illustrate the two types of Third World ENGOs. In Costa Rica, for example, the Association of New Alchemists (ANAI) is a professional local ENGO which supports grassroots organisations that are involved in the promotion of conservation-development initiatives in areas threatened by environmental degradation. The organisation was founded in 1971 by North American and Costa Rican scientists and today is run by a team of professional local ecologists, agricultural extentionists, agronomists and community representatives. Projects initiated and supported by ANAI often involve agro-forestry schemes which provide local grassroots actors (e.g. poor farmers) with timber from sustainably managed forests for their own use in an attempt to halt local deforestation (Carroll, 1992; Fisher, 1993; Utting, 1993). In contrast, *Fundación Natura* in Ecuador is a national middle-class advocacy organisation that campaigns on a variety of issues, such as national park development and biodiversity protection, dear to that country's growing middle class. This ENGO is staffed mainly by trained professionals, and works in close association with First World ENGOs such as the WWF and the American Nature Conservancy, notably over debt-for-nature swaps (Meyer, 1995; Bailey, 1996; Wapner, 1996).

The preceding discussion has sought to give a general sense of the potential political significance and heterogeneity of ENGOs as they relate to the Third World context. Not surprisingly, the diversity of ENGOs is linked to the growing differentiation of civil society itself in the Third World. Thus, the combination of democratisation and uneven economic growth in selected countries has prompted the development of a middle-class environmentalism that finds its expression, in part, through the growth of ENGOs that reflect middle-class concerns about environmental quality. The political implications of such diversity will become apparent below, and in the next chapter, when a detailed analysis is conducted. For our purposes, the following discussion will be structured loosely around the typology of ENGOs presented in Table 6.1 – that is, grassroots organisations; grassroots support organisations (or service organisations), and regional, national and international advocacy organisations. It should be noted that this is a general schema that does not purport to cover all ENGOs in the diversity of their makeup or activities.

Table 6.1 Main types of ENGOs

Grassroots Organisations (GOs)	These organisations are based within urban or rural communities where community members have organised themselves into a movement to campaign about particular environmental issues which have a direct impact upon community quality of life. In the Third World, environmental GOs often develop as a response to threatened livelihood concerns, such as the depletion of forest resources, upon which local incomes depend, by logging and mining practices. However, many also develop from anxieties concerning general quality of life within the community, such as pollution from nearby factories or from a lack of adequate basic amenity provision in the community.
Grassroots Support Organisations (GRSOs)/Service Organisations (SOs)	These organisations are often established by teams of highly skilled professionals, such as scientists, economists, agronomists or ecologists, in both the First and Third Worlds. The intent is either to help support community-based movements in the promotion of environmentally sustainable projects, or by themselves to initiate such projects in areas threatened by ecological destruction from local economic practices. Unlike GOs, many GRSOs/SOs are non-membership based, and are staffed by educated non-community members.
Regional, National and International Advocacy Organisations (AOs)	These are often large, highly professional organisations which initiate campaigns concerning a multitude of environmental issues at the regional, national or international scale. Campaign methods often involve publicity-grabbing educational programmes and the lobbying of governments, multilateral institutions or businesses on a variety of environmental concerns such as tropical deforestation and nuclear arms testing. The most prominent AOs are the membership-based First World NGOs such as Friends of the Earth and Greenpeace. However, many First World-established AOs do have a direct influence in the Third World for a number of reasons. First, they are made up of affiliations, many of which are now based in the Third World. Second, many of the issues they campaign on, are directly relevant to the Third World, for example, deforestation. Increasingly, indigenous national and regional AOs are being established within the Third World, particularly in those regions with a growing middle class (e.g. South and South-East Asia and Latin America).

Source: Adapted from Clark, 1991: 40–1

The rest of this chapter focuses on formal 'professional' ENGOs – that is, regional, national and international advocacy organisations on the one hand, and grassroots support organisations on the other hand. Chapter 7, in turn, concentrates on the role of informal 'non-professional' grassroots organisations that seek to represent directly the interests of poor grassroots actors.

FIRST WORLD ADVOCACY ENVIRONMENTAL NGOs

Like the transnational businesses with which they often come into conflict, successful First World advocacy ENGOs have become global enterprises in that they maintain multi-country operations and tackle a wide range of environmental problems. The relationship between these ENGOs, the specific problems they choose to highlight in their campaigns, and media coverage of those campaigns is complex (Princen and Finger, 1994). What needs to be emphasised here is that Third World environmental problems – especially tropical deforestation, biodiversity loss and the elimination of 'big' wildlife – have become major preoccupations of First World ENGOs. Not only have organisations such as FOE, Greenpeace or the WWF become known around the world in the process, they have also served to emphasise certain aspects of the Third World's environmental crisis over other aspects. This highly selective approach has been severely criticised by some Third World ENGOs as representing a distortion of the Third World's 'most pressing problems' (see below).

To understand the role of First World ENGOs in the Third World's environmental crisis is to appreciate the origins and development of this group of organisations. As noted, these ENGOs developed in the First World during the social and environmental protests of the 1960s and 1970s. The initial concern of these organisations was to focus on issues relating to endangered wildlife (e.g. whaling, the killing of Arctic seals) and industrial pollution of most immediate concern to their First World supporters (Yap, 1989/90; Stoett, 1993; McCormick, 1995). Such environmental activism reflected a wider public anxiety in the First World about the environmental implications of industrialisation, which was fuelled by the work of influential natural scientists like Carson (1962) and Commoner (1972). Combining publicity-grabbing environmental activism with (sometimes) carefully prepared scientific briefs, the leading First World ENGOs pressured states and businesses in the First World to alter environmentally degrading practices and, by the late 1970s, organisations such as Greenpeace and FOE had already largely outgrown the radical image of a decade earlier.

Two trends in the development of First World ENGOs in the 1980s are of particular interest here. First, these actors began to take a growing interest in 'global' environmental problems such as greenhouse warming, ozone depletion or tropical deforestation that led them to take a direct interest in environmental problems in the Third World. First World ENGOs had certainly expressed some interest in 'Third World' problems prior to the 1980s, notably at the United Nations Conference on the Human Environment held in Stockholm in 1972 (McCormick, 1995). However, this interest had been tangential to their main concern with those environmental problems perceived to be of most relevance to the First World. These were related to the destructive effects of intensive agro-chemical use, polluting industries and the loss of wildlife habitats. Yet, beginning in the 1980s, First World ENGOs began to develop system-atically a Third World dimension to their campaigns such that today those campaigns are as likely to focus on tropical deforestation or the illegal ivory trade as they are to relate to industrial activity or road-building in the First World (Livernash, 1992; Bramble and Porter, 1992).

The incorporation of Third World environmental problems into these campaigns has nonetheless been highly selective, reflecting a decision to give priority to those problems that resonate well with the First World public (see below). Indeed, the disproportionate attention that has been given to the problems of wildlife depletion and tropical deforestation in the campaigns of many First World ENGOs reflects a long-standing First World fascination with the Third World's tropical forests and exotic wildlife (Putz and Holbrook, 1988; Hecht and Cockburn, 1989; Beinart and Coates, 1995).

Second, First World ENGOs became much larger and more professional in their activities and outlook in the 1980s, and these changes have been reflected in their Third World campaigns and political influence. The evolution of these organisations was partly a response to the maturation of their members and activists, from a young, and largely student body in the 1960s, to a thirty- or forty-something middle-class group in the 1980s. That evolution was also linked to the very success of these organisations in the 1960s and 1970s. Many First World ENGOs increased rapidly in size, budget capacity and 'global reach' in the late 1970s and 1980s, as offices and/or affiliates were set up around the world, including the Third World. Greenpeace, for example, grew from

> a single office in Vancouver [Canada, in 1972] to staffing offices in over thirty countries and, until recently, a base in Antarctica. Greenpeace has offices in the developed as well as the developing world ... Its eco-navy consists of eight ships, and it owns a helicopter and a hot-air balloon. It employs over 1,000 full-time staff members, plus hundreds of part-timers and thousands of volunteers. As of July 1994, it had over 6 million members worldwide and an estimated income of over $100 million. Finally, it has expanded its area of concern. While originally focused on nuclear weapons testing, it is now concerned with all threats to the planetary ecosystem.
>
> (Wapner, 1996: 47)

The WWF and FOE also became multi-million-member organisations controlling multi-million-dollar budgets in the 1980s with extensive representation in such countries as Brazil, India and Malaysia. Such international growth brought these First World ENGOs into ever closer contact, and often conflict, with states, transnational corporations (TNCs) and multilateral institutions around the world. This trend enhanced their political position generally as they became major players at the local, regional and even global scale (Bramble and Porter, 1992; Wapner, 1995).

First World ENGOs have used their growing political power worldwide to mount high-profile advocacy campaigns designed to alter the policies and practices of various powerful actors. Thus, ENGOs have taken states around the world to task over their 'abdication' of the state's stewardship role (see Chapter 3). Various strategies have been tried in the process, but ENGOs often coordinate their campaigns in a bid to pool resources and maximise political effect (although a lack of cooperation among ENGOs can also be a problem, see below). A common tactic has been to target First World states as part of an attempt to prod these states into placing pressure on Third World states (via trade restrictions or aid and loan disbursements) with the ultimate aim of effecting policy change in the Third World itself.

A case in point is the campaign by a number of First World ENGOs to ban the ivory trade in Africa. The campaign developed in the 1980s at the behest of the WWF, Trade Records Analysis of Flora and Fauna in Commerce (TRAFFIC), and the Environmental Investigation Agency (EIA), following alarm expressed in a number of African countries over rapidly decreasing elephant numbers (Princen, 1994b). Despite the inclusion of the African elephant in 1976 on the list of endangered species regulated under the Convention on International Trade in Endangered Species of Wild Fauna and Flora (CITES), African states made little effort to end the ivory trade, with the result that elephant numbers continued to decline rapidly. However, in the 1980s, First World ENGOs campaigned vigorously on the issue, lobbying in particular the United States Congress for a complete ban on the trade. Success in this regard was reflected in 1988 in the passage of the United States African Elephant Conservation Act, a law which required the Secretary of the Interior to investigate all ivory-producing states in order to determine which had effective elephant management schemes in place. The following year, many First World states (and the European Union) declared a moratorium on ivory imports, and four African ivory-producing nations proposed a total ban on the trade (Princen, 1994b).

First World ENGOs have also mounted highly effective campaigns against the policies and practices of multilateral institutions with links to Third World environmental problems, once again focusing their energies on those First World states with the most influence over these institutions. The most prominent example of such a campaign to date has been that which was conducted against key international financial institutions in the United States in the mid-1980s. This campaign was initiated by North American

ENGOs led by the Environmental Policy Institute and the National Wildlife Federation. It was part of an attempt to stop multilateral institutions, such as the World Bank or the Inter-American Development Bank, from providing cheap loans to Brazil and other Third World states, which, in turn, would facilitate policies perceived to be bad for both Third World peoples and environments (Bramble and Porter, 1992; Rich, 1994). Environmental NGOs focused on lobbying the United States government since this actor is the leading source of funds for the multilateral banks noted above and, accordingly, has the largest say on their executive boards. This campaign was successful in that, as a result of unrelenting pressure on senior US politicians in Congress and in the administration, the ENGOs persuaded the United States government in March 1985 to vote against a half-billion dollar power sector loan proposed by the World Bank on environmental grounds – a move that prompted a policy 're-think' by the Bank soon after (see Chapter 4).

First World ENGOs also seek to persuade businesses, especially TNCs, to alter their environmental practices. Chapter 5 notes that businesses have been major sources of environmental degradation in the Third World, and yet retain strong influence in national and international policy-making circles. Although they are no match for much larger and better financed TNCs, First World ENGOs have nonetheless been quite skilful in forcing the hand of these economic Leviathans by mounting campaigns designed to increase popular awareness about the environmentally degrading practices of targeted businesses. Such campaigns are notably associated with calls for the public to boycott the products of 'eco-hostile' businesses, and it is the prospect of a loss of custom that has led even the largest TNCs to back down in the face of such ENGO campaigns (Wapner, 1995).

To take but one example, the United Kingdom branch of FOE has instigated a range of public awareness/boycott campaigns targeting selected business practices with the aim of altering or even stopping them. Thus, it has conducted a campaign to end the trade in mahogany, a valuable commercial timber extracted primarily from Brazil's biologically diverse Amazon rainforests, in a bid to stop the widespread deforestation and social genocide associated with that trade (Secrett, 1987; Juniper, 1992). In a hard-hitting 'Mahogany is Murder' campaign, this ENGO has promoted popular awareness in the United Kingdom of the social and ecological pitfalls of the mahogany trade, and has urged a consumer boycott of all mahogany products sold in furniture and 'do-it-yourself' stores so as to render the trade unprofitable in that country (Friends of the Earth, 1993). At the same time, it has sought to encourage an alternative 'rainforest harvest' that would protect the forests and the people living in those forests from extinction (Friends of the Earth, 1992). FOE (United Kingdom) has also attacked Rio Tinto Zinc, one of the world's largest mining TNCs (see Chapter 5), over that corporation's plans to mine in Madagascar, a biologically rich island off the south-eastern coast of Africa, which has already seen many of its species eliminated as a result of the expansion of cash-cropping and other economic activities (Jarosz, 1993; Moody, 1996).

These various campaigns have often led states, businesses and multilateral institutions to modify policies and practices linked to the Third World's environmental crisis. Yet First World ENGOs have been criticised themselves, notably by groups in the Third World who suggest that these organisations pay inadequate attention to the livelihood dimensions of that crisis. Many First World ENGOs, it is asserted, have pursued an 'environment-first' policy rather than a 'people-first' policy in the Third World. In some cases, this may have been to the detriment of poor grassroots actors.

Two examples illustrate this point. The campaign to ban the ivory trade noted above may have led to a diminished worldwide trade in ivory products, as well as to an expanded effort by African states to protect elephant populations, but it has also resulted in growing resentment among farmers and pastoralists in the affected areas. These grassroots actors have found that the protection of the African elephant championed by selected First World ENGOs has often led to their own exclusion from areas set aside to protect prized elephant populations. For those not excluded, increasing elephant numbers have been a major nuisance, in that wandering elephants have destroyed crops vital to local livelihoods. As Princen (1994b:151) observes, policies to protect the African elephant have engendered 'local resentment among resident peoples which easily translates into increased poaching and encroachment on protected lands'. Indeed, and returning to a theme first developed in Chapter 3, these policies have resulted in 'coercive conservation' (Peluso, 1993b), that is, in policies in which states seek to impose their will on recalcitrant or hostile local populations.

Debt-for-nature swaps provide a second example of an 'environment-first' policy that has often worked to the detriment of local actors. These swaps were devised in the mid-1980s to address two problems simultaneously: Third World debt and the Third World's environmental crisis (Cartwright, 1989). Following the example of the United States' Conservation International, which effected the first major swap in Bolivia in 1987, a number of First World ENGOs have followed suit in such countries as Costa Rica and Madagascar. The procedure involved is that part of a Third World country's debt is paid off by an ENGO, in return for which the state in question agrees to set aside a specified ecologically sensitive area for permanent protection. Yet many of these debt-for-nature swaps are instigated without due regard for the livelihood needs of the populations living in the designated areas. Indeed, since the schemes usually involve a policy of complete protection, those grassroots actors whose livelihoods are dependent on the exploitation of natural resources contained within those areas are usually severely affected. The result is all but inevitable. Farmers, shifting cultivators or other grassroots actors are denied access as part of the terms of the swap, and resort thereafter to illegal forest extraction or land uses (e.g. grazing). State officials respond with arrests and local intimidation, and the cycle of conflict is established (Hayter, 1989; Mahoney, 1992).

In Colombia, for example, a debt-for-nature swap has resulted in the creation of a protected area in a remote but biologically important inland forest. Yet this scheme, which received the strong support of several First World ENGOs, has garnered considerable opposition from local grassroots actors. The latter's access to natural resources, upon which they depend to earn a livelihood, has been threatened, without even the offer of adequate compensation for the loss of income.

First World ENGOs are becoming sensitive to the charge that their campaigns may inadvertently prejudice the livelihoods of grassroots actors. Many are now trying to link conservation to local development concerns in a bid to provide these actors with alternative 'sustainable' livelihoods. For example, recent WWF initiatives seek to stimulate community participation in eco-tourism ventures, agro-forestry schemes and horticulture projects. Yet even these well-intentioned efforts can backfire when they do not come to grips with local concerns. In Thailand, for instance, the WWF has worked on a series of projects in villages surrounding the Khao Yai National Park. Local farmers, however, feel that they have benefited very little from these projects since they have tended to concentrate on education, rather than on the income-generating activities of most immediate concern to these individuals (Ghimire, 1994).

The role of First World ENGOs in the attempted resolution of the Third World's environmental problems is thus far from straightforward. They have certainly played an important role in fighting the environmentally degrading practices of traditionally powerful actors. Their dedication to promoting the 'environmental cause' has won them much public admiration and support, especially in the First World, and they have increasingly sought to throw their weight behind integrated conservation-development projects in the Third World (Barrett and Arcese, 1995; Bailey, 1996). First World ENGOs nonetheless stand accused of having tended to view the Third World's environmental crisis in a highly selective fashion, focusing on problems, such as elephant depletion and tropical deforestation, of interest mainly to their First World supporters. Further, the 'success' of some of their environmental campaigns has seemingly been predicated on the denial of the local rights of certain grassroots actors. However, First World ENGOs are beginning to realise that an approach in which environmental concerns are divorced from local livelihood issues is unlikely to result in either social well-being or environmental conservation.

A potentially more serious criticism of First World ENGOs centres on the 'reformist' character of most of these organisations. Thus, even when they do address social issues, they tend to advocate reforms designed only to alter incrementally the political and economic status quo. The policies and practices of states, businesses and multilateral institutions should be changed only so far as is required to guarantee sustainable development under the present capitalist system (e.g. IUCN *et al.*, 1991). In the case of 'conservative' First World ENGOs, close cooperation with these actors is the order of the day to the extent that a number of these organisations are supported financially by their erstwhile opponents. Chatterjee and Finger (1994: 109) note that in the 1980s 'it became quite acceptable for environmental NGOs to solicit donations from private corporations ... by the end of the 1980s most Northern NGOs had levered substantial corporate contributions. Many of them even had joint programmes with corporations.' They cite the case of the Environmental Defense Fund (EDF), one of the United States' biggest ENGOs, which has signed two major contracts with McDonald's and General Motors to cooperate on environmental matters. Although both of these corporations 'agreed not to use the agreement for publicity and both allowed EDF publicly to criticize their policies', each TNC has obtained 'free mileage' out of the deal through the ability to 'portray themselves as compromisers and listeners' without changing appreciably their business practices (Chatterjee and Finger, 1994: 109–10). The EDF, in turn, has little to show environmentally for its cooperation with business – the professed reason for the deal, after all.

The issues raised by these flourishing links between First World ENGOs and traditionally powerful actors are serious in that they potentially call into question the legitimacy of these organisations as 'dispassionate defenders' of environmental concerns. Indeed, if the argument is that fundamental change is the only way in which to resolve the Third World's environmental crisis, then such behaviour means that many First World ENGOs may be 'part of the problem, not part of the solution' (Middleton *et al.*, 1993). In contrast, ENGOs indigenous to the Third World may be able to avoid the difficulties of their First World counterparts, and remain focused on a livelihood-based approach to the Third World's environmental crisis. This issue is examined in the next part of this chapter.

THIRD WORLD ENVIRONMENTAL NGOs

First World ENGOs are important players in efforts to deal with the Third World's environmental crisis, but their Third World counterparts are also fast acquiring an influential role in this process. As with First World ENGOs, most of these Third World ENGOs are drawn from the middle classes, and tend to reflect middle-class concerns about the deteriorating quality of the environment. Third World ENGOs are usually staffed with trained professionals whose primary objectives are to lobby states, businesses or multilateral institutions about their environmental policies and practices; to protect ecologically sensitive areas; and to instigate sustainable development programmes. As in the First World, these organisations vary considerably in size, structure, range of activities and ideology. Third World ENGOs are closely linked in many cases to their First World counterparts, who often provide them with technical and financial support, a situation that leaves them open to the criticism that they are 'tools' of First World interests (Eccleston, 1996). Yet other Third World ENGOs are relatively autonomous and possess considerable power in their own right (Fisher, 1993).

There are thousands of professional ENGOs in the Third World today, but their geographical distribution is far from equal. Thus, there is a high concentration of these organisations in South and South-East Asia, as well as in Latin America. In India, for example, there are well over 500 professional ENGOs while in the Philippines, the figure may be as high as 1,000 (Rush, 1991; Broad, 1993). Similarly, Latin America may encompass as many as 6,000 professional ENGOs (Branes, 1991; Price, 1994; Meyer, 1995). In contrast, there are very few indigenous professional ENGOs in Africa. One reason for this situation may relate to the arrested development of a middle class in most (but not all) countries on this continent, reiterating the general point made earlier about the association between the emergence of professional ENGOs and the existence of a well-developed middle class. Another reason may be linked to the persistence in many parts of Africa of authoritarian states which tend to repress autonomous activity (see below).

However, the rise of the professional Third World ENGO is a relatively recent phenomenon, certainly when compared with the emergence of its counterpart in the First World. It is only really since the late 1970s that the Third World has seen the development of indigenous professional organisations geared specifically to campaigning on environmental issues. In India, for example, more than 50 per cent of India's professional environmental NGOs were established as recently as the 1980s, with 81 per cent in total established since 1970 (Rush, 1991). There are two possible reasons as to why this is so.

First, intensifying environmental problems linked to uneven economic development began to cause anxiety among the Third World's emerging middle class at the same time as the tenets of First World popular environmentalism first became widely known to members of this class, notably through the auspices of First World ENGOs (see above). Specifically, the recognition has grown among members of the middle class that industrialisation is not the unmitigated good they had long thought it to be. Rather, industrialisation is increasingly seen as a process responsible for intense urban air, land and water pollution (the main site of industrial production), and for accelerating rural forest loss and land degradation (due to resource extraction for industry) (McDowell, 1989; Rush, 1991).

That the adverse environmental consequences of industrialisation have long been recognised by poor grassroots actors directly affected by such development is a theme explored elsewhere in this book (Chapters 2 and 7).

Yet such degradation has only come to be perceived as such relatively recently by the more affluent middle class, which, it must be said, owes its prosperity largely to economic activities that often created such degradation in the first place. Along with growing affluence has nonetheless come a desire for a 'cleaner' urban environment as well as a rural environment selectively protected for the 'consumption' of the Third World's middle class (e.g. national parks). It is this concern that underlies the rapid development of the professional Third World ENGO.

Second, the growth of an 'environmentally conscious' middle class in the Third World is also linked to the advent of democratic or 'quasi-democratic' political regimes in many parts of the Third World since the 1980s. To the extent that civil society as a whole has won a larger voice in national political and economic affairs, the concern of private citizens over declining environmental conditions has been publicly expressed, often via organisations that criticise and lobby the state openly. A number of Third World professional ENGOs were certainly formed in the 1960s and 1970s, but these organisations typically had to maintain a low profile during years of authoritarian rule. In the Philippines, for example, a number of professional ENGOs were established in the 1970s under the authoritarian regime of President Ferdinand Marcos, but wisely eschewed political activism until the overthrow of that regime in February 1986. Thus, the country's leading professional ENGO, Haribon Foundation, was established in 1972 but confined its attention to the non-political issue of the protection of endangered birds for the duration of the Marcos era. Thereafter, this organisation, and many others like it, took a more active political role in the nation's environmental affairs encouraged by the democrat President Corazon Aquino herself in this regard (Broad, 1993; Vitug, 1993). A similar process has occurred in the case of professional ENGOs located in other countries recently converted to democratic rule, notably Brazil, Thailand and South Korea (Hirsch and Lohmann, 1989; Silva, 1994; You, 1995).

The relationship between democracy and the spread of professional Third World ENGOs (and associated ENGO networks) nonetheless defies simple explanation. Malaysia, for example, boasts a long-standing democratic regime and a burgeoning ENGO movement. Yet the freedom of the latter to criticise the policies and practices of the Malaysian state was curtailed in the late 1980s as a result of the growing irritation of Prime Minister Mahathir with the 'rash' behaviour of organisations like Sahabat Alam Malaysia and the World Rainforest Movement (see below, and Chapter 7).

In contrast, Indonesia is ruled by the authoritarian Suharto regime, yet this regime has recently experienced increasing difficulty in constraining ENGOs, notably those affiliated to the influential Indonesian Forum for the Environment (WALHI) (Eccleston and Potter, 1996). Although WALHI has maintained close links with the Indonesian state since the former was established in 1980, it has not been afraid 'to embarrass the government internationally', or, for that matter, to take specific state agencies to court over alleged environmental malpractices (Eldridge, 1995: 137). Yet the dismal experience of ENGOs in such countries as Vietnam or Burma still ruled by authoritarian regimes suggests nonetheless the existence of a link between democratisation, the development of a middle class and the rise of professional Third World ENGOs (Rush, 1991; Eccleston and Potter, 1996).

What seems clear is that these organisations have played a growing political role wherever they have been able to develop. However, the Third World professional ENGO often differs from its First World counterpart in that the former, although representing middle class concerns, has been more likely than the latter also to defend the interests of poor grassroots actors. This quest to represent simultaneously middle-class and grassroots interests reflects the widespread tendency in much of the Third World to associate environmental problems with basic livelihood issues, given the existence of ubiquitous poverty in the region (Redclift, 1992). Whilst by no means necessarily opposing capitalism, many professional ENGOs in the Third World recognise that capitalist development has had serious adverse effects upon both the environment and many of the people most intimately linked to that environment. That recognition has led these organisations to instigate campaigns that not only seek to protect the environment, but that also aim to address the livelihood concerns of poor grassroots actors. While most Third World professional ENGOs seek to integrate development and environmental concerns in their work, differences nonetheless exist among them that arise from differing approaches to those concerns, and even from the ways in which the ENGOs themselves are organised. This point may be illustrated with reference to the advocacy ENGOs and grassroots support ENGOs referred to above.

Advocacy ENGOs are perhaps the most widely known of these two types of ENGO as a result of high-profile national and international campaigns on behalf of poor grassroots actors adversely affected by the development process. These ENGOs do not typically conduct work at the grassroots level themselves, but rather provide financial and technical support to grassroots organisations in order to sustain the latter's struggles with state agencies, multilateral institutions or businesses (see Chapter 7). Their role is above all to coordinate campaigns and publicity so as to enhance the prospects for the success of grassroots actors in these struggles. As such, these ENGOs are seen, and often see themselves, as political agents seeking to change the political and economic status quo.

Advocacy ENGOs have had a major impact on environmental conflict in various national settings. A case in point is one of Thailand's largest advocacy ENGOs, Project for Ecological Recovery (PER). Founded only in the mid-1980s, PER quickly became the country's central coordinating forum for a diverse coalition of grassroots actors, middle-class students, professionals and politicians disenchanted with the Thai state's environmental policies and practices (Leungaramsri and Rajesh, 1992; Permpong-sacharoen, 1992). Although this ENGO has been involved in diverse lobbying and public education campaigns, it came of age in the battle over the proposed Nam Choan dam when it spearheaded a nationwide struggle to prevent this dam's construction in the mid- to late 1980s. The proposed dam was to be built in an area of outstanding natural beauty in the north-west of the country encompassed within the Thung Yai Naresuan Wildlife Sanctuary, and was also to have an adverse impact on thousands of poor local farmers, many of whose homes and land were to be flooded to create the dam's reservoir (Hirsch and Lohmann, 1989; Rigg, 1991). Based in the nation's capital, Bangkok, PER coordinated the protests of various grassroots and middle-class groups; it also 'acted as an information clearing house, led reporters to "good" stories, and kept an open line to people in government' (Rush, 1991: 75). With such media coverage, the campaign also gained international attention, and many First World ENGOs added their support to the cause. By early 1988, the campaign had reached such proportions that the Thai government decided to postpone the project indefinitely, thereby handing a victory to the PER-led coalition.

Yet the high-profile nature of much of the work that advocacy ENGOs do may also be a potential hindrance to the cause of grassroots activism. Thus, these organisations may unintentionally stiffen the resolve of states wishing not to be seen to 'give in' to 'radical' groups. The experience of Malaysia's Sahabat Alam Malaysia (SAM) over the struggle of the Penan illustrates this point. This organisation was established in 1977 as an affiliate of FOE, but has maintained an autonomous stance vis-à-vis its First World 'parent' organisation (Eccleston and Potter, 1996). As with Thailand's PER, SAM has campaigned on a wide variety of national environmental concerns and is also involved in nationwide environmental education initiatives designed to promote a cleaner environment. Like PER it has also become embroiled in national controversy over its campaigns on behalf of disadvantaged grassroots actors, with the hunter-gatherer Penan being the most well-known case. As noted in Chapter 7, the latter have been fighting a bitter struggle against the local government in the east Malaysian State of Sarawak, as well as the loggers closely linked with that government, who are together destroying the forests upon which the Penan depend (Hong, 1987; Colchester, 1993). Largely as a result of SAM's intervention (acting in conjunction with the World Rainforest Movement), the plight of the Penan has become a *cause célèbre* in environmental circles worldwide, and both the local Sarawak government and the federal Malaysian state headed by Prime Minister Mahathir have been inundated over the years by petitions calling for logging to be halted. This issue has even been raised at the diplomatic level, and has become a major source of international embarrassment for the Mahathir government (WRM and SAM, 1989). Yet, if anything, SAM's national and international efforts have only reinforced the resolve of both

147

levels of government to perpetuate the status quo. Thus, logging has persisted despite the protests and the status of the Penan has continued to decline. Further, SAM and other ENGOs have felt the ire of the Mahathir government directly as their ability to operate freely in the country has been restricted in various ways over the years (see above).

The modus vivendi of Third World advocacy ENGOs thus constitutes a high-risk strategy that aims in many cases directly to challenge the political and economic status quo. Yet it is far from clear how effective this political strategy is, or even whether these organisations are best placed to provide the practical, and often highly location-specific, assistance that many grassroots actors seem to require. Indeed, and notwithstanding their professed goal of integrating environment and development considerations, there is still a tendency on the part of many Third World advocacy ENGOs (like their First World counterparts) to favour environmental conservation over local livelihood concerns 'when push comes to shove'. This comment applies not so much to 'radical' ENGOs like PER or SAM, but rather to 'conservative' organisations that mainly seek to protect threatened habitats or species and who, as a result, tend not to incorporate local actors' concerns and participation systematically into their initiatives. To take but one example, *Fundación Natura* in Ecuador has initiated in recent years agro-forestry and horticulture schemes in communities living close to national parks in a bid to stop the exploitation of natural resources found in these parks by local farmers (Bailey, 1996). However, the results of these initiatives have been mixed, with some projects working well both socially and environmentally, but with other projects failing mainly because they did not provide viable economic alternatives to local actors. A common reason for failure is that the projects (particularly the agro-forestry schemes) seem to be geared more to replenishing the forests than to providing local actors with satisfactory livelihoods. A related problem is the lack of the local actors' participation in the design and management of the projects, which is held to explain the relative neglect of the livelihoods issue (Bailey, 1996).

It might be expected that grassroots-support ENGOs are better placed by their very nature to avoid the potential weaknesses of advocacy ENGOs. Thus, the former usually eschew high-profile policy advocacy in favour of a low-key 'problem-solving' approach in which attention is focused on environmental problems at the local scale, rather than on broader political or economic issues. Grassroots-support ENGOs tend to be less confrontational in their method of action than is the case with advocacy organisations. However, the essentially political role of the former should not be overlooked for, in their own way, they are representative of the way in which social groups in many Third World countries have begun to intervene in economic and social processes hitherto largely the preserve of powerful actors (Friedmann, 1992). The results of such intervention may be ultimately more dramatic than those attained by advocacy ENGOs in so far as they demonstrate the viability of alternative development strategies (Broad, 1993).

Indeed, grassroots-support ENGOs appear the more effective of the two types of professional Third World ENGOs in terms both of reaching the poorest members of society and of integrating conservation and development concerns. This may be because grassroots-support ENGOs are usually set up with specific aims in mind that are devised in close consultation with grassroots actors themselves. ANAI, noted briefly above, is a case in point. This ENGO operates in Costa Rica's Talamanca canton and has developed various agro-forestry projects in the region to supply poor grassroots actors with an alternative to destroying the forests (Carroll, 1992). These actors play a major role in devising and managing ANAI projects so they are able to shape the projects to suit their own needs. Indeed, they are well represented in the ENGO's management structure itself, and therefore play a key role in the overall formulation of ANAI policies. Some 6,000 people have benefited directly from ANAI's agro-forestry projects, with one-and-a-half million trees already planted, enhancing the livelihood prospects of all the actors that participate in the projects (Carroll, 1992).

Not all grassroots-support ENGOs succeed in this manner. Indeed, many fail to live up to the high expectations placed on them due to inadequate funding and the related inability to 'scale up' successful projects or to provide participants with tangible economic benefits (Farrington and Bebbington, 1993). Perhaps most surprising of all is the apparent lack of community participation in the design, development and management of the projects of numerous grassroots-support ENGOs despite an adherence to this ideal in principle. Vivian (1994) notes that many such ENGOs in Zimbabwe promote themselves as highly participatory in nature, but in practice rarely reflect the concerns and wishes of grassroots actors. Livernash (1992: 223) reports that in a study of seventy-five projects set up by grassroots-support ENGOs in various parts of the Third World, it was discovered 'that what was often termed participation was in practice a form of decentralised decision-making still dominated by NGO staff and local elites, and that local elites received a disproportionate share of benefits'. In these situations, grassroots-support ENGOs fail to assist, and may even hinder, the most marginal actors in society who are supposed to be the main project beneficiaries. To a certain extent, the problem may reside in the vicissitudes of funding available to these ENGOs who must often rely heavily on First World ENGOs and other actors for their survival (see below).

While many Third World professional ENGOs attempt to integrate environmental and development concerns in their work, this goal is often not achieved with the result that the livelihood problems of poor actors remain unaddressed. In the case of Third World professional ENGOs still attached to an 'environment-first' approach, it would seem as if these two concerns were incompatible. As Adams (1990: 190–1) argues, 'the primary objectives of conservationists are protected areas and species. Development, even if packaged as "sustainable development", is attractive chiefly as a secondary strategy where it promotes their primary objective.'

ENVIRONMENTAL NGOs AND THE STATE

To some extent, Third World professional ENGOs put their relative inability to respond effectively to the plight of poor actors down to 'interference' by the state. Many ENGOs have tended to view their relations with states in terms of oppression and confrontation. As Rush (1991: 88) argues, 'a common feature of the environmental movement everywhere' is that it has been 'poised against "government"'. This perception is scarcely surprising since it is the state which controls many key decisions concerning the environment, and thus it is the state which ultimately is to blame when environmental degradation and social injustice prevail (see Chapter 3). Campaigns by Third World professional ENGOs thus challenge the role of the state as the nation's 'steward' and typically assert grassroots actors as the 'real' environmental managers.

The discussion so far has suggested that an essentially adversarial relationship exists between ENGOs and the state. In some cases, an ENGO is apparently able to prompt a state to change its policies in just such a context. Thus PER successfully pressured the Thai state to abandon the planned Nam Choan dam, while in western India, the Bombay Environmental Action Group (BEAG, formed in 1977 by a group of middle-class professionals) managed to persuade the state to alter its decision to construct a fertiliser and petrochemical plant near Bombay (Rush, 1991; Permpongsacharoen, 1992). In other cases, an ENGO appears powerless to effect any change in the activities of the state. The inability of either SAM in Malaysia or WALHI in Indonesia to stop their respective states from continuing to support destructive logging in tropical rainforest are two examples (see above).

Yet this image of ENGOs in perpetual conflict with the state is only part of the story since ENGO campaigns in many parts of the Third World are predicated on cooperation with the state. In Latin America, for example, official state support for ENGOs varies from country to country – in Brazil, the importance of ENGOs is actually enshrined in environmental legislation which stipulates that the National Environmental Council (CONAMA) must include ENGO representatives (Branes, 1991). In the two cases noted above in which ENGOs were successful, both PER and BEAG relied extensively on contacts within the state to increase the pressure on political leaders, a move which assisted these organisations to turn the tables on their opponents. Thus, 'quiet' diplomacy between ENGO members and state officials complemented the often harsher public stances taken by the protagonists.

Indeed, many states make it a regular practice actively to support 'moderate' factions within the ENGO community. In India, for example, Rush (1991) reports that a number of the country's leading professional ENGOs, including the world-renowned Centre for Science and Environment (CSE), receive financial and technical support from the state. The CSE, an influential think-tank cum environmental-advocacy organisation, has received considerable official funding to gather and publish data on the links between economic activities, and the everyday and episodic environmental changes that might be connected to those activities (Centre for Science and Environment, 1985; and Chapter 2).

States also establish working partnerships with ENGOs on specific environmental campaigns and projects. In the Philippines, for example, the ENGO community is a major vehicle for the state's 'community forestry' policy with ENGOs serving as a critical link between the Department of Environment and Natural Resources and the People's (i.e. grassroots) Organisations responsible for policy implementation at the local level (Braganza, 1996). States may also look to ENGOs for new development ideas or technical expertise. Farrington and Bebbington (1993: 153) report that in Kenya the Forestry Extension Services Division has established a strong relationship with some of the country's leading ENGOs so as 'to draw on their experience to stimulate participatory approaches to extension and service provision'. States may even go so far as to create their own ENGOs – stretching to the limit the meaning of an environmental 'non-government' organisation in the process. In Costa Rica, the state developed the Neotrópica Foundation and the National Parks Foundation in order to facilitate the creation of national parks and ecological reserves (Livernash, 1992). These organisations were set up to buy land in ecologically sensitive areas to be designated as parks or reserves and play a crucial role in the management of those protected areas. A similar role is played by the many 'GRINGOs' (government run/initiated NGOs) operating today in the Philippines which provide invaluable assistance to the state (Constantino-David, 1995).

The evidence on ENGO–state relations is thus mixed. In some cases, these actors become locked in confrontation, while in other cases they cooperate in the pursuit of common goals. Cooperative relations illustrate in particular that many ENGOs realise in practice that to succeed with their campaigns they must rely to some degree on the state – their frequent criticisms of this actor notwithstanding. As Clark (1991: 75) notes, ENGOs 'can oppose the state, complement it, or reform it – but they cannot ignore it'. This acknowledgement of the necessity of the state places many ENGOs in a quandary. On the one hand, cooperation with the state provides funds and security that might otherwise be lacking – with important implications for the fate of ENGO campaigns. The provision of funds means that the power and influence of ENGOs may increase in society, thereby facilitating the success of their campaigns. Conversely, a lack of state support may mean that ENGO initiatives may be consigned to insignificance (Clark, 1991; Sanyal, 1994).

Further, the ability of ENGOs to operate at all is often dependent on the permission of the state. This point is fairly evident in relation to those countries in which autocratic rule is still the norm. In countries such as Indonesia, Zimbabwe or Zaire, ENGOs are only tolerated so long as they meet with the approval of state leaders. In this regard, states can use their law-making powers to keep 'troublesome' ENGOs in check. In Indonesia, for example, under the ORMAS (Social Organisation) Law of 1986, ENGOs must register with the Ministry of Home Affairs, and all sources of ENGO funding must be approved by the state; the law even permits the state to ban any ENGO without explanation (Sinaga, 1994). However, even in 'democratic' countries, the state may use its legal powers to intimidate what it considers to be 'hostile' ENGOs. As noted above, in Malaysia the controversy over the plight of the Penan prompted the Mahathir government to tighten the rules that govern ENGO operations as part of a seemingly successful campaign to bring 'unruly' organisations to heel.

On the other hand, cooperation with the state may weaken the credibility and power of ENGOs. Such cooperation may erode the autonomy, and even undermine the purpose of what, after all, is called a *non-government* organisation. It was suggested earlier that the power and prestige of the ENGO has been derived in part from the sense of legitimacy and impartiality associated with this type of actor's separation from traditionally powerful actors like states, businesses or multilateral financial institutions. To the extent that ENGOs conduct their work in conjunction with, or even on behalf of, the state, for example, they run the real risk of forgoing the popular goodwill that has been the bedrock of their power over the years. As Sanyal (1994: 37) notes, while the state was seen as the 'bad guy', ENGOs were able to revel in their role of 'good guy' but if ENGOs were to side with the state, 'eventually they would be either controlled or coopted by the state, thereby losing their legitimacy and effectiveness'.

Indeed, in their often desperate scramble for funds, many ENGOs have become increasingly dependent on the state for support, with the clear *quid pro quo* being the understanding that ENGOs will not seek to challenge the power and authority of the state (Yap, 1989/90). States are thus in a potentially powerful position vis-à-vis ENGOs. However, state control of ENGOs need not be as crude as the threat to withdraw legal entitlements or funds (Smillie, 1995). Thus, states often coopt members of ENGOs on to official boards and commissions in a move which may increase ENGO influence but is more likely only to 'dull the sharp edge of NGO criticism and occupy the attention of much of the best NGO talent' (Clark, 1991: 79). Even worse, by working with the state in this manner, ENGOs may simply be strengthening the state's overall position in society in so far as the public message conveyed is that incremental political reform is the 'best' way to address the Third World's environmental crisis.

The ability of most ENGOs to resist cooption by the state must certainly not be under-estimated. The prospect of funding from First World ENGOs or aid agencies, not to mention the support of grassroots actors themselves, reduces to some extent the dangers that ENGOs will simply become the handmaidens of the state. Yet, the power of the state in these matters must also be recognised, if only as a necessary antidote to the enthusiasm of some proponents of the powers of ENGOs (e.g. Ekins, 1992). The rise of ENGOs may be an important factor in the reconfiguration of power relations and environmental conflicts in many parts of the Third World today, but they are nonetheless only one actor among others in political-ecological struggles. It would be most unwise therefore to exaggerate their importance (Smillie, 1995).

A NEW POLITICS OF THE ENVIRONMENT?

The question still needs to be asked as to whether the advent of ENGOs signals the start of a new politics of the environment. It may be the case that it is too early to answer that question – although the possible significance of ENGOs for environmental conflicts at the grassroots level is assessed in the next chapter. Here, it is useful briefly to consider relations within the ENGO community to try to ascertain whether a new style of politics is reflected in the way that ENGOs relate to each other.

The evidence on this score is mixed. There are already a considerable number of ENGOs in existence in the Third World but, despite the commonly held belief that ENGOs are 'models of co-operation' (Sanyal 1994: 41), in many cases, these organisations operate alone rather than in a partnership with other organisations. To some extent, it is plausible to assume that such isolation stems from the peculiarities of the business – that is, the need to tailor operations to location-specific circumstances reduces the prospect for fruitful collaboration between ENGOs. However, the tendency to 'go it alone' is also related to the increasingly competitive nature of that business which often sparks off intense rivalries between organisations. Such rivalries may be linked to various causes, but funding is often a leading one.

That scarce funds often drive Third World ENGOs into the arms of the state has already been noted. However, even for ENGOs that cooperate with the state, funding is tight, and for the vast majority of these organisations a hand-to-mouth existence is the norm. The all but inevitable result is that ENGOs must compete for scarce funds, and one measure of the 'prowess' of these organisations is their ability to win the support of key funding bodies – viz., First World ENGOs, First/Third World state agencies, businesses or multilateral institutions. The extent to which ENGOs tailor their environmental campaigns to the interests of potential or actual sponsors, and the different strategies that they use in order to raise funds, is a relatively under-explored subject (Meyer, 1995; Smillie, 1995). However, if the experience of First World ENGOs is anything to go by, then the tendency to tone down criticism of potential backers may become the norm, if it has not already become so in many places.

It would nonetheless be wrong to imply that the sum total of relations within the ENGO community is that of a 'war of all against all'. Rather, and recognising that isolation and competition often only result in increased political and economic vulnerability as well as insecurity for both the ENGOs and the grassroots actors they aim to assist, there has been a recognition that ENGOs may all benefit by selectively pooling their technical and financial resources in a common effort (Livernash, 1992; Chatterjee and Finger, 1994; Sanyal, 1994). Indeed, the networking of ENGOs through special federations is now well under way.

It is perhaps a reflection of the enduring political importance of the state (see Chapter 3) that many such federations take the form of national associations or fora that are designed explicitly, in the first instance, to lobby the state. In Indonesia, for example, more than 400 environmental organisations are linked together through WALHI. This umbrella organisation serves, among other things, to give smaller and less powerful ENGOs a more effective voice in efforts to lobby the Indonesian state, and seeks in general to put forward a common ENGO perspective on official policies. Thus, WALHI played a crucial role in the 1980s in helping a number of the country's ENGOs to bring the country's first lawsuit against a business (allied to the Suharto government) which had violated national environmental laws (Livernash, 1992; Eldridge, 1995; and see above).

These national federations also provide a forum in which ENGOs can share their expertise and knowledge with each other, so that projects and campaigns can be fine-tuned or even replicated elsewhere. In the Philippines, for example, diverse ENGOs use the Green Forum-Philippines, a loose federation created in early 1990, as a means of giving each other support, exchanging expertise and working together on various sustainable development campaigns and projects (Legazpi, 1994). Perhaps the most important function of national federations such as Green Forum-Philippines or WALHI-Indonesia, however, is that, through regular conferences, they help to build a 'country-wide dialogue' among ENGOs, thereby 'linking local issues to national issues and local actions to national actions' (Rush, 1991: 81).

In a similar fashion, ENGOs have joined together to form regional networks in a move designed to take ENGO perspectives to regional fora. Thus, the development of organisations such as the Asia-Pacific People's Environmental Network (APPEN), founded in 1983, and ANEN (African NGOs Environmental Network), started in 1986, signal the determination of the ENGO community to make its voice heard regionally, just as an active

ENGO presence at international environmental summits, such as the Rio Earth Summit of 1992, reveals a quest to present a case at the global decision-making level (Chatterjee and Finger, 1994). At times, these national, regional or global initiatives are deceptive in that they appear to give the picture of an ENGO community in accord. Yet, none of these fora have been able to overcome the rivalries (notably over funding) noted above. Indeed, in some respects, they may have served to magnify such rivalries (Legazpi, 1994).

A further potential source of tension within the ENGO community concerns the relationship between First World and Third World professional ENGOs. The main bone of contention here is the dependence of the latter on the former for funds. Although in many respects a 'model of cooperation', the relationship between the two is nonetheless based on unequal power relations in a manner strongly reminiscent of the relationship between Third World and First World states (see Chapter 3). However, if cooperation rather than conflict has been the norm in the ENGO community, there are signs of growing tension between First World and Third World ENGOs. Thus, Livernash (1992: 228) suggests that 'many Southern groups believe that their dependence on Northern groups for funds has exacted too high a price to pay in [terms of] lost autonomy, compromised priorities and reduced institutionalised identity'. Many Third World ENGOs are now seeking greater autonomy from their First World counterparts: 'as Southern NGOs have raised their numbers, skills and management capability, they are increasingly eager to set the agenda on their own terms' (Livernash, 1992: 228). Many Third World ENGOs are thus, for example, emphasising that, while First World ENGOs may possess the technical, economic and scientific expertise with which to launch large-scale environmental campaigns in the Third World, organisations indigenous to the Third World are better placed to understand the needs, concerns and issues of Third World communities caught up in environmental conflict. As a result, it is argued, there needs to be a transfer of responsibility in dealing with the Third World's environmental crisis from the First World to the Third World. The fact that First World ENGOs have had a tendency to adopt an 'environment-first' approach over the years has only reinforced the argument that Third World ENGOs are best left in charge of coordinating the ENGO community's response to Third World environmental concerns (although, as we have seen, some Third World ENGOs have been equally guilty of neglecting livelihood issues – see also Meyer, 1995).

Such a transfer of responsibility may already be under way. Livernash (1992) notes that some First World ENGOs are establishing 'partnerships' with their Third World counterparts in which the former focus on providing technical and economic support but leave the latter to deploy those resources more or less as they see fit. In Indonesia, for example, two Canadian ENGOs have established a working partnership with three Indonesian organisations.

The Canadian organisations transfer technical and scientific know-how to their Indonesian counterparts to help them in an environmental appraisal of hydrological systems, but it is left to the Indonesians to decide how to go about the process itself (Livernash, 1992). Related to this tendency is another trend noted above – namely, that First World ENGOs are beginning to learn from their counterparts in the Third World the critical need to abandon an 'environment-first' approach in favour of an integrated approach in which livelihood issues predominate.

The tendency for the ENGO community to be dominated by the interests and concerns of First World organisations has nonetheless yet to disappear. At the 1992 Earth Summit, for example, Third World ENGOs found that the ENGO agenda was largely set in keeping with the interests of their powerful First World counterparts. As Chatterjee and Finger note, from the perspective of many Third World ENGOs,

> at least on paper the Rio framework was totally in line with their own approach to environment and development. However, in practice, concrete participation in the UNCED process turned out to be quite difficult for them. One of their major problems was the lack of organisation. As a result, many Northern conservationist NGOs functioned as a voice of Southern NGOs.
>
> (Chatterjee and Finger, 1994: 76)

Third World interests were overshadowed by those of First World ENGOs in the process and, as with the inter-state forum itself, environmental considerations often seemed to prevail over development concerns. There is a long way to go, therefore, before the relationship between Third World and First World ENGOs approaches the equality upon which it has, theoretically at least, always been based. And if unequal power relations can be seen to be so evident even within the ENGO community, what hope is there for a new politics of the environment based on the ENGO example?

Nevertheless, the dramatic growth in the size, power and influence of this new type of actor in both the First and Third Worlds, whose political significance has been the subject of this chapter, is not to be gainsaid. Its significance can be measured in at least two ways. It can be measured in terms of the ability of ENGOs to force traditionally powerful actors – states, businesses, multilateral institutions – to modify those practices that are associated with the Third World's environmental crisis. As noted, various campaigns by First World and Third World ENGOs have managed to force a change in the policies or practices of actors contributing to Third World environmental degradation (see above and Chapter 4). The ability to expose environmental malpractices, draw media (and hence public) attention to the contradictions between the policy pronouncements and practices of various actors, organise boycotts, or assist in the empowerment of grassroots actors, all serve to illustrate the ways in which ENGOs become embroiled in political-ecological conflicts, and, in doing so, the possible ways in which they may alter the topography of a politicised environment.

It is nonetheless striking how little the prevailing unequal distribution of power between actors has changed. Notwithstanding altered policies and practices here and there, the basic forces linked to the global capitalist system, which many political ecologists feel is at the heart of the Third World's environmental crisis, remain largely intact. This situation was nowhere more evident than at the 1992 Rio Earth Summit where the outcome was shaped, as usual, by those actors – states, multilateral institutions, TNCs – who have long dominated environmental management issues (see Chapters 3–5). This outcome occurred despite the presence of a large number of ENGOs at Rio, who in many cases made their opposition well known to the assembled world media. As Chatterjee and Finger note,

> the UNCED documents were hardly affected by the various NGOs. This is with the exception of some mainstream environmental NGOs, whose positions were so close to the positions of the governments, that their distinctive impact can hardly be detected in the texts.
>
> (Chatterjee and Finger, 1994: 98)

Rather than showing the power of the ENGO community, Rio appeared only to reveal its fundamental weakness in comparison with more powerful actors. It seemed that the only way that ENGOs could claim an input on the proceedings was by effectively validating what had been decided already by states, multilateral institutions or businesses. Thus, the ENGO community 'did not emerge from Rio stronger, but weaker. As a result, it is more fragmented and more disoriented than before' (Chatterjee and Finger, 1994: 65–6).

However, the political significance of ENGOs can also be measured in another way, with reference to the effect that they may have on the livelihood struggles of grassroots actors. This chapter has suggested that professional ENGOs – especially, but not exclusively, from the First World – have experienced considerable difficulties in integrating environmental and development considerations in their activities, but that this situation is changing as a result of pressure from their Third World counterparts (a number of whom have also been guilty of a similar 'offence'). Whether this change continues may well be critical to the future of environmental conflict and change in the Third World, in that it promises to pose a challenge to traditionally powerful actors from a direction which they fear the most – namely, the grassroots. It is to a consideration of the prospect for fundamental change 'from below', and the grassroots organisations and actors that would be associated with such a change, that this book now turns.

7

GRASSROOTS ACTORS

A central theme in Third World political ecology since its inception has been the political and ecological oppression of grassroots actors by more powerful actors such as states or businesses. The ability of grassroots actors, a category which variously includes shifting cultivators, small-scale farmers, nomadic pastoralists, hunter-gatherers, poor urban dwellers or fishers, to resist the predations of the powerful is also emphasised in the work of political ecologists. Yet as various chapters in this book have shown, grassroots actors have more often than not been at the losing end of environmental struggles with their lot in a politicised environment one largely characterised by marginality and vulnerability.

This chapter examines the role of grassroots actors in such an environment, and considers the socio-economic impact on these actors of restricted access to environmental resources. Grassroots actors rarely accept this fate passively (although we give illustrations of 'adaptation'), but strike out at their oppressors in both covert and overt ways. Such resistance is not new as the discussion of grassroots resistance during the colonial era illustrates. Much of this resistance took the form of covert activities since public opposition usually incurred a highly repressive response from colonial states – a response, moreover, routinely resorted to by many postcolonial Third World states as well. However, the recent spread of democratic or quasi-democratic political regimes in the Third World has reduced the likelihood that states will behave in this manner. An intriguing, and potentially revolutionary, development in terms of the topography of a politicised environment since the 1980s has thus been the emergence of grassroots organisations as a political force to be reckoned with in many parts of the Third World. This chapter investigates the meaning of these organisations as part of a wider assessment of the prospects for, and possible political significance of, the empowerment of grassroots actors.

ACCESS, LIVELIHOODS AND ENCLOSURE

That the environment in the Third World is largely a livelihood issue has been noted at various stages in this book. The growth of 'First-World-style' environmentalism among members of the Third World's prospering middle class has certainly been associated with intensifying calls for environmental conservation based on aesthetic reasons (see Chapter 6). It is nonetheless the case that virtually everywhere in the Third World today, livelihood concerns remain a central issue in understanding the political implications of the environmental crisis.

As political ecologists suggest, this crisis relates to the question of livelihoods which, in turn, is linked in a very specific manner to the unequal power relations discussed in Chapter 2. We note in that chapter how powerful actors (e.g. states, businesses) may derive power from an ability to control the environmental resources of weaker grassroots actors (e.g. poor fishers, shifting cultivators), and how the latter become marginalised and especially vulnerable to environmental degradation. It is useful here to begin the discussion of grassroots actors by spelling out more fully the livelihood implications of this process, and the grassroots management practices that are thereby disrupted.

The link between the livelihoods of poor grassroots actors and environmental change may be seen in both urban and rural areas in the Third World. In the former, many of the livelihood activities of the urban poor concentrate heavily on the exploitation of surrounding natural resources. Thus, nearby rivers and coastal waters provide fish or crabs, while local forests and scrublands provide timber or fuelwood, for domestic consumption or sale. Yet as forests are cleared or water courses are polluted, the ability to derive a livelihood in this manner is reduced, if not eliminated altogether (Hardoy *et al.*, 1992).

However, it is perhaps in rural areas that the links between environmental change and grassroots livelihood concerns are most evident. For grassroots actors living in these areas, 'survival in the short term is their primary concern and for this they depend largely on the resources of the surrounding area' (Elliott, 1994: 63). On the one hand, the combination of extreme poverty and limited or non-existent access to fertile agricultural land, plus other environmental resources, results in few opportunities for poor grassroots actors to escape their dependent circumstances simply through hard work (in many cases, these actors are already working with maximum effort). On the other hand, the existence of few other economic opportunities in many rural areas of the Third World often forecloses the pursuit of alternative livelihoods by these actors. Seasonal out-migration is certainly one way in which grassroots actors may seek to make ends meet (Blaikie, 1988), and rural opportunities for the poor are not invariably dire (cf. Rigg, 1997, on South-East Asia). It is nonetheless the case that in many rural areas, circumscribed livelihood prospects and ubiquitous poverty go hand in hand. These conditions render the link between the physical environment and grassroots livelihoods all the more significant since the opportunity for poor grassroots actors simply to move to new areas if local environmental resources are degraded or depleted is often virtually non-existent. The existence of new 'frontiers' for potential migrants is rapidly becoming a thing of the past in most parts of the Third World.

An important implication of this situation is that it is usually in the interest of poor grassroots actors to manage environmental resources in a sustainable manner, not so much because these actors necessarily have a greater 'respect' for the environment, but rather because their livelihoods depend on the maintenance of those resources in a way that is usually not the case with more powerful actors. Redclift notes,

> To most poor people in rural areas, for whom daily contact with the environment is taken for granted, it is difficult, if not impossible to separate the management of production from the management of the environment, and both form part of the livelihood strategy of a household or group.
>
> (Redclift, 1992: 36)

That livelihood strategy may be associated with a sophisticated under-standing of the location of potential environmental resources and the ways in which local ecological processes operate. It is often also embodied in the development of complex institutional arrangements regulating the use and management of resources (especially common property resources).

The whole question of 'local' or 'traditional' environmental knowledge is the subject of long-standing debate and controversy among policy-makers and scholars. During the colonial era, the state belittled such knowledge as 'primitive' and 'inefficient' as part of a broader attempt by the coloniser to control commercially valuable environmental resources within the conquered territories (see Chapter 3). This attitude has persisted in official circles even to present times, although perhaps with decreasing policy significance as it is also increasingly recognised that indigenous knowledge is neither primitive nor necessarily inefficient (Chambers, 1983; Richards, 1985; Murdoch and Clark, 1994).

That knowledge is often as diverse as the social and ecological contexts within which it has developed. For example, fishing communities living in the Marituba wetlands of northern Brazil have used a number of strategies to ensure that only adult fish are caught so as to maintain fish supplies over the long term. Tree branches baited with cassava are laid in the water at selected points where the fish are to be found, but such bait is only attractive to the adult fish, thereby protecting young fish from premature capture (Diegues, 1992). Similarly, many Karen shifting cultivators in southern Burma (Myanmar) have long attempted to make their annual forest clearances for temporary agriculture (*taungya*) in such a way as to allow recently used lands the necessary time to recuperate before further exploitation occurs. A complex system combining rotational forest clearance and fallow has been the means used to guarantee itinerant agriculture in the area (Bryant, 1994a). There are many other examples of how indigenous knowledge has been applied so as to promote long-term environmental management goals – indeed, the work of political ecologists is rife with them (e.g. Richards, 1985; Little and Horowitz, 1987; Hecht *et al.*, 1988; Gadgil and Guha, 1992). The general point here is that indigenous knowledge usually reflects a detailed appreciation and understanding of local environmental resources by grassroots actors, and that such knowledge has often served as the basis for highly effective environmental management systems allowing for simultaneous resource exploitation and conservation.

The livelihood quest of grassroots actors has also been reflected in the development of local institutional arrangements to regulate use by individuals or groups of environmental resources. Such arrangements can be quite complex and may involve multiple and overlapping user rights to land, water, trees, animals or other resources (Berry, 1989; Blaikie, 1989; Peters, 1994). The management of common property resources has been especially important yet complicated, and has often resulted in the development of highly complex common property regimes (CPRs) that bring together different individuals in a community with the aim to manage those resources on a communal long-term basis (Ostrom, 1990). Such CPRs typically involve two things: well-defined common property resources and resource users; and strict controls on access and use so as to prevent over-exploitation. A number of important points follow from this understanding of CPRs. First, 'open-access' resource use is not permitted since strict controls are placed on resource use by all actors in the community, and external actors are usually excluded. As Bromley (1991: 25) argues, 'common property represents private property for the group of co-owners (since all others are excluded from use and decision making)'. Second, CPRs are based on local control and decision-making, enabling the adjustment of management practices in the light of altered social and environmental conditions. Third, CPRs are never fixed in space or time but fluctuate depending on shifting property relations, environmental circumstances, and social conflicts among actors both inside and outside a CPR (*Ecologist*, 1993). Fourth, CPRs have provided a flexible institutional means by which grassroots actors have often sought to reconcile resource use with environmental conservation – although the existence of a CPR in itself has never been a guarantee of success in this regard (Ostrom, 1990).

There are numerous examples of CPRs in the literature, but two examples will have to suffice here (see Ostrom, 1990, for other examples). The semi-nomadic pastoralist Barabaig people of Tanzania have long sought to regulate communal use of such common property resources as rangeland, water or trees through an hierarchical system of institutions: community-wide public assemblies for men (*getabaraku*) and women (*girgwageda gademg*), neighbourhood councils (*girgwageda gisjeud*) and clan councils (*hulandosht*). Through these institutions, the Barabaig have recognised that, 'to make efficient use of resources, access [to environmental resources] needs to be controlled to prevent exploitation beyond their capacity to recover' (Lane, 1992: 89). The second example relates to the irrigation societies or *zanjeras* of Ilocos Norte in the northern Philippines, of which there are between 1,000 and 1,200 in existence today (some in operation for up to two centuries). These organisations are established to build and manage irrigation systems on behalf of local communities so as to enable a regular supply of water to fields for agricultural production. The *zanjeras* have needed to allocate and mediate water rights or shares (*atar*) among community members in such a way as to ensure that all actors receive an equitable share of available water supplies at the required times (i.e. in keeping with crop needs) but also that those supplies are not depleted through over-exploitation. The organisational complexity of these organisa-tions varies with simpler *zanjeras* headed by only a leader (*panglakayen*) and a secretary/treasurer, while more complex societies have a much larger management team often plus a board of directors. However complex, the benefits of this system of community management frequently include 'productivity and equity' gains, which also 'contribute to the system's environmental sustainability' (Yabes, 1992: 119).

The development of CPRs is one way in which grassroots actors have long sought to regulate resource use through specific management practices. To appreciate their existence and role is also to begin to clarify one of the big debates that has surrounded the environmental management practices of many grassroots actors over the years – namely, whether the Third World's growing environmental crisis reflects a 'tragedy of the commons' (Hardin, 1968) or a 'tragedy of enclosure' (*Ecologist*, 1993).

Hardin's thesis on the tragedy of the commons was explored in Chapter 3 in the context of a discussion of the theoretical justification of the state. Here it is useful simply to point out that the 'flip-side' of that justification was a critique of the ways in which grassroots actors managed – or more accurately 'did not manage' – commons resources. Yet even the brief discussion above on indigenous knowledge and CPRs should make clear that the reality of grassroots environmental management has usually been a far cry from the abstract description provided by Hardin. Thus, 'far from being a "free-for-all", use of the commons is closely regulated through communal rules and practices' (*Ecologist*, 1993: 13). As many scholars point out, what Hardin is actually describing is an 'open-access' situation whereby resources are open to use by all actors, and there are no CPRs or other management structures to regulate such use (Blaikie and Brookfield, 1987; McCay and Acheson, 1987; Berkes, 1989; Ostrom, 1990; Bromley, 1991; Peters, 1994).

Indeed, research by political ecologists has been instrumental in pointing out that the Third World's environmental crisis reflects mainly a tragedy of enclosure rather than a tragedy of the commons (*Ecologist*, 1993). In this process, the state, often acting in conjunction with businesses and multilateral institutions, denies grassroots actors access to commons resources hitherto managed by them through local institutions such as CPRs. In effect, CPRs are taken over by the state for large-scale commercial exploitation either by its own agencies or by allied business interests using the legal-political powers of the state. A notable case in point has been the creation of extensive networks of reserved forests, national parks and 'government lands' in many parts of the Third World (Guha, 1989; Peluso, 1992, 1993b; Utting, 1993; Bryant, 1994b). The habitually exploitative practices – styled 'development' in the postcolonial era – carried out in these 'nationalised' territories has included notably large-scale logging, mining, cattle ranching, cash-crop production and dam construction. The ways in which traditionally powerful actors have promoted, and benefited from, these activities has been a recurring theme in this book. Here, the implication of the enclosure of the commons for poor grassroots actors requires discussion.

The first thing to note in this regard is that the enclosure of the commons was typically associated with the dissolution of many of those grassroots institutional arrangements (notably CPRs) that had hitherto managed the commons. As power over local environmental resources shifted from grassroots actors to the state and other actors external to the community (e.g. national firms or transnational corporations – TNCs), the need for these grassroots institutions largely disappeared and, with it, the utility of local cooperation in aid of long-term environmental management. To be sure, not all of these institutions disappeared, and local public 'passivity' in the face of outside management of local resources often belied fierce conflict over access usually conducted by covert means (see below). However, a corollary of 'development' has been undoubtedly the weakening, if not the elimination altogether, of grassroots environmental management in much of the Third World.

The second point to note is that the enclosure of the commons served to further marginalise poor grassroots actors in the measure that their access to environmental resources essential for their livelihoods was restricted or denied. Not only was access to commons resources ended, but these actors were often forced into a situation whereby they had to work ecologically marginal lands elsewhere in order to survive. The concurrent decline of grassroots institutions meant that there were few, if any, avenues open to individuals seeking redress. The end result was that marginalisation became a defining trait for most poor grassroots actors as they were displaced from newly created reserved forests, national parks or other 'development' projects.

To take but one example, the construction of large dams has resulted in the displacement of hundreds of thousands of poor farmers, hunter-gatherers or shifting cultivators, as their homes have been submerged to create huge reservoirs (Goldsmith and Hildyard, 1984, 1986; Cummings, 1990). In India, the state has sought to dam the Narmada river in the north-west of the country since the 1970s primarily in order to meet India's growing energy needs fuelled, in turn, by the development process. Although this project – one of the largest such projects in the world (30 major dams + 135 medium dams + 3,000 minor dams) – will benefit businesses as well as some urban dwellers and agriculturalists, controversy surrounding the project has centred on the planned displacement of over 200,000 mainly poor people living in the path of the planned development (Rush, 1991). The state has certainly offered to resettle these people, but the indicated locations are generally much less fertile than those areas from which they have been evicted. Further, many of these grassroots actors are tribal groups who do not wish to leave ancestral lands. For all of these reasons, this project has been dubbed 'the world's greatest environmental disaster' (Alvares in Gadgil and Guha, 1994: 112) and has prompted concerted local, national and international protest.

Yet to speak of the marginalisation of poor grassroots actors is also to acknowledge that some actors within this broad category have been worse affected than others by this process. Poor women and indigenous minorities in particular have apparently borne a disproportionate share of the costs associated with such marginality. It is useful here to explore briefly why this is so, as well as to consider what the ensuing consequences have been.

To appreciate why poor women are often the most severely affected by the Third World's environmental crisis is first to understand the argument that most poor women in the Third World have a closer relationship with the environment than most poor men (Shiva, 1988; Sontheimer, 1991; Agarwal, 1992; Jackson, 1993; Mies and Shiva, 1993). However, there is considerable debate in the 'feminist political ecology' literature as to why this situation is the case. On the one hand, an 'essentialist' argument is put forward by the likes of Shiva (1988), and Mies and Shiva (1993) in which women's closeness to the environment is related directly to the reproductive role of women in society. Thus, the ability of women to reproduce and create life, to rear and nurture the young, is similar to the ability of surrounding ecological processes to reproduce and create life. On the other hand, a 'materialist' viewpoint is asserted by such scholars as Agarwal (1992) and Jackson (1993) in which the intimate relationship between women and the environment is related squarely to the material needs and position in society of women. To be sure, an important part of Shiva's (1988) argument is that women are materially dependent on environmental resources, and that their tendency to promote sustainable environmental management practices is above all related to this situation. Yet Agarwal (1992: 125) argues that Shiva is insufficiently materialist in that this writer 'does not differentiate between women of different classes, castes, races, ecological zones, and so on'. By lumping together all women vis-à-vis the environment, Agarwal suggests that Shiva fails to focus adequately on the plight of poor women whose dependence upon environmental resources is much heavier than is the case for wealthier and politically powerful women. Indeed, Shiva's approach suggests a false unity of interests that, for example, is unable to account for why some women choose to degrade the environment (cf. Jackson, 1993).

Be that as it may, this debate has served to emphasise the plight of poor women whose livelihood strategies often rely heavily on the exploitation of resources to provide fuel, fodder, food and forms of income for themselves and other family members. Examples of the strong dependency of poor women on the surrounding environment are numerous, but one example from Sierra Leone will suffice here to illustrate the point. Leach (1993) describes the reliance of poor women belonging to the Mende tribe upon local plant resources. The Mende are swamp rice farmers living on the edge of Sierra Leone's tropical forest zone. The exploitation and dependence upon plant and forest resources is nonetheless much more apparent amongst women than it is amongst men. While men grow the staple crop (rice), the women must provide alternative staple supplies from the forest in the period before the rice harvest. They must also supply vegetables, fish and meat even when the rice is available. Further, while men often control key income-generating activities, such as cocoa and coffee farms, women have few comparable opportunities. As such, poor women depend almost entirely upon the forests to supplement their meagre incomes, notably through handicrafts production. Yet income derived in this manner is small compared with that frequently earned by men from the more lucrative activities noted above.

As a result of such dependency, poor women have been especially hard hit by the combined effects of the enclosure of the commons and associated environmental degradation. In many cases, the working day of poor women has increased dramatically as they have had to walk further and further to collect badly needed resources such as water and fuelwood (Shiva, 1988). In rural South Asia, for example, domestic fuelwood is almost always collected by women, yet this essential activity has become an ever more laborious and time-consuming task as the effects of massive deforestation take their toll. In parts of Nepal, where fuelwood collection took women only a couple of hours twenty years ago, it now takes them most of the day. The marginalisation of poor women has also increased within the household and local community as their ability to earn a separate income is reduced, if not eliminated altogether (Joekes et al., 1995). To some extent, this situation reflects the additional time that women must spend each day doing other chores (as above). However, it also reflects the loss of those environmental resources needed to earn an income as a result of enclosure and/or environmental degradation. Thus, forests have long provided poor women with the raw materials with which to make handicrafts, and thereby to earn a small independent income. This activity is difficult due to deforestation in many parts of the Third World today. To these factors must also be added the possibility that women's health is often imperilled as a result of enclosure and environmental degradation. The reduction in the quantity or quality of water and fuelwood supplies can have especially dramatic consequences for the diet of poor women. With less fuel, for example, they are able to cook less food, and what meagre portions are cooked are usually offered first to husbands and children, resulting in under-nourished women. Indeed, this unequal distribution means that these women 'are unlikely to get the extra food necessary to make up for the additional energy they expend in fuel collection' (Agarwal, 1988: 100). Poor women in this way bear the brunt of the costs associated with the marginalisation of poor grassroots actors as a whole.

As for poor women, the combined impact of enclosure and environmental degradation on indigenous minorities in the Third World has been massive as a result of their generally heavy reliance on diverse environmental resources. Many such groups have not even today been fully incorporated into the global capitalist system, but rather seek to perpetuate livelihood practices at the margin of that system which are intimately associated with careful environmental management (Hong, 1987; Denslow and Padoch, 1988; however, this separation is not always the case, see Doedens et al., 1995). To be sure, it is important not to 'romanticise' the practices of these actors as templates of sustainability (Murdoch and Clark, 1994). Research by political ecologists has nonetheless built up a picture of the close links that usually exist between indigenous groups and environmental resources as a result of the dependency of these actors on the surrounding environment for food, fuel, shelter and other basic needs (Hong, 1987; Anderson and Huber, 1988; Hecht and Cockburn, 1989; Saldana, 1990; Blauert and Guidi, 1992). Indeed, the sheer breadth of the knowledge of these actors is often breathtaking: the Shuar Indians in Amazonian Ecuador use 800 species of plants and tree products for medicinal, food, fodder, fuel and shelter purposes, for example. The combination of such detailed environmental knowledge, and the existence in many cases of robust community environmental management structures, has been at the root of suggestions that 'sustainable use of local resources is simple self-preservation for people whose way of life is tied to the fertility and natural abundance of the land' (Durning, 1993: 91).

As a result, the marginalisation of indigenous groups has been especially catastrophic, and these actors face the unhappy distinction of being 'the most marginal of the marginal'. Causes of environmental destruction in indigenous areas are numerous, but as Durning (1993) notes, because many indigenous groups live in forested areas, commercial logging is a major culprit. Thus, Hong describes how the livelihood of the Penan hunter-gatherers of Sarawak has been undermined by logging in that Malaysian state:

Wild sago (their staple diet) is being destroyed by the felling and the wild boar and other animals are being frightened away by the noise of tractors and chainsaws. Timbermen also shoot these animals and use *tuba* [the roots of a local liane] to poison fish, depriving the Penan of their food.

(Hong, 1987: 90)

Such logging not only destroys Penan livelihoods, it also calls into question what it means to be 'Penan': 'their culture is fast being eroded and their identity as a community is also being threatened' (Hong, 1987: 235). A similar process of social and economic marginalisation is taking place with reference to non-forest-dwelling indigenous peoples as well. In Tanzania, for example, extensive rangelands used by the semi-nomadic pastoralist Barabaig people have been 'enclosed' by the state, and legal title to these lands transferred to large commercial farming businesses supportive of the Tanzanian state. In the process, the Barabaig have been excluded from lands that they have long managed carefully (see above). As they have needed to pursue a livelihood from a smaller territory, the Barabaig have become 'unwilling parties to the destruction of the land' (Lane, 1992: 93). That livelihood has also been squeezed as grazing cattle has become more difficult, and cattle numbers have plummeted: in 1988, for instance, Barabaig cattle herds in some areas were one-third of their 1981 level due to 'mortalities attributed to the stress of reduced grazing' (Lane, 1992: 96–7). The political and economic pressures that have marginalised the Barabaig people further intensified in the early 1990s as the post-socialist Tanzanian government has pursued IMF-inspired structural adjustment policies (Neumann, 1995). As with the Penan, this Tanzanian indigenous group is thus being subjected to political and economic forces that threaten the very cultural survival of the group itself.

The loss of control over homelands through resource enclosure or environmental degradation typically leads to the disintegration of indigenous management systems that may have been an effective means of long-term resource use. Indeed, as noted above, indigenous actors may even be forced into a situation in which they must join in the degradation of the local environment on the principle that 'if you can't beat them, join them' (Blaikie and Brookfield, 1987). In Amazonian Ecuador, for example, Bebbington *et al.* (1993) note that the Shuar Indians have been forced to take-up economic practices such as cattle ranching, which require extensive destruction of local forests in a manner similar to colonists, in a desperate bid to secure rights to their land. If they did not clear the forests in this way, the Shuar might lose their land to colonists able to claim the 'undeveloped' forest land as their own under Ecuadorian law. Yet, even if the Shuar are able to retain control of their lands in this manner, they have nonetheless had to degrade their own environment and surrender a cherished way of life in the process.

167

The preceding discussion has sought to outline the relationship between questions of access, livelihoods and enclosure. The emphasis has been on poor grassroots actors who have been most adversely affected as a category, although the especially acute plight of poor women and indigenous minority groups was also highlighted. What emerges from this discussion is the strong dependency of poor grassroots actors on the environment on the one hand, and their growing social and ecological marginalisation as a result of the tragedy of enclosure on the other hand. Yet traditionally powerful actors have not had an easy time of it as they have sought to displace grassroots actors from control over local environments. Indeed, and as this chapter will later show, the latter have mounted a campaign of resistance through public grassroots organisations, which in many parts of the Third World today is beginning to reverse the process of enclosure perpetrated by powerful state and non-state actors. First, however, it is useful to explore the contexts in which open resistance is not offered either because poor grassroots actors have opted to adapt to the altered political and economic situation, or because, through fear of reprisal, they have chosen to resist in a covert manner.

ADAPTATION AND EVERYDAY RESISTANCE

Poor grassroots actors have long responded to enclosure and related environmental degradation in ways which are designed to maintain livelihood opportunities, but which also avoid provoking powerful actors into any retaliatory action that might exacerbate their plight.

A strategy of adaptation has been used frequently by these actors over the years, especially in countries subject to authoritarian rule. However, even under conditions of democratic rule, the political and economic marginality of these actors is such that it is often impossible for them to protest about the environmental degradation or physical exclusion that is disrupting their livelihoods. Many grassroots actors nonetheless adopt strategies that aim to minimise any adverse effects on them while at the same time avoiding confrontation with powerful actors. For example, poor grassroots actors adapt to enclosure or environmental degradation by extending the time spent pursuing livelihood needs. The cost to these actors in this case is borne through a heavier workload. The time spent gathering fuelwood and water is thus prolonged, especially, as noted above, for poor women (Agarwal, 1988; Shiva, 1988). The result – longer working days devoted to basic chores and consequently less time for income-generating activities or leisure – represents a classic and ubiquitous case of adaptation.

A further adaptive response is to utilise diverse social and economic 'coping' strategies which may originate with 'traditional' responses to naturally occurring environmental processes (i.e. drought), but which are also deployed in the context of enclosure or environmental degradation. Such strategies notably encompass the modification of economic practices, the storage of crops from good seasons, the sale of livestock or the request of assistance (usually labour) from neighbours and relatives (Mortimore, 1989; Elliott, 1994). However, as political ecologists note, these coping strategies have become less effective in the measure that the processes of enclosure and environmental degradation have intensified (Watts, 1983a; Richards, 1985; *Ecologist*, 1993).

In a few cases, adaptation may be associated with a partial reversal in the marginalisation of poor grassroots actors as these actors take advantage of new economic opportunities generated by the capitalistic market. Thus, for example, Amanor (1994) reports that some grassroots actors living in the forest and savannah frontier zones of Ghana use their local environmental knowledge to develop new economic strategies to reverse environmental degradation in these areas. Fuelwood shortages, for instance, have been partly rectified by the planting of fast-growing exotic trees in place of the slower-growing indigenous trees; these provide for local needs and afford an additional income through market sales. Similarly, new crops have been introduced, and diversified cropping patterns pioneered, to try to combat the problem of declining soil fertility. Thus, early maturing cassava has now replaced traditionally grown types which are longer maturing, while cowpea has replaced maize in the minor season to cope better with reduced rainfall and soil fertility. Economic diversification is also occurring in livestock production and in the development of a local oil-palm distilling industry, both of which represent alternative income for those actors burdened with limited or heavily degraded land. This example illustrates the general point that poor grassroots actors are not necessarily helpless in the face of broader political, economic or ecological changes. Indeed, they may even benefit from such changes provided suitable 'resourcefulness' is shown (Rigg, 1993). Yet the ability to fight the marginalisation process through 'entrepreneurial' skill alone should not be exaggerated if only because the economic strategies resorted to may lead to the depletion of essential local resources and, ultimately, the intensified impoverishment of the grassroots actors in question.

This point is clearly illustrated in cases where grassroots actors feel that they have no alternative but to partake in the degradation of their local environment. The case of the Shuar Indians of Amazonian Ecuador (above) has already been referred to in this context. Perhaps the best example of such 'cooptation' is when forest-dwellers throw in their lot with the state-sponsored loggers who are destroying their forests in a desperate bid to earn a livelihood. A notable case in point is the decision of some members of the Dayak community in Sarawak to join Chinese logging firms acting in conjunction with the Malay-controlled Sarawak state (and often financed by Japanese TNCs) in the felling of the region's hardwood forests. These Dayaks find employment as loggers or truck drivers in the forest industry or, in the case of politically important village leaders, may even have been directly 'bought off with timber concessions' (King, 1993: 243). Such recruitment is actively encouraged by logging promoters since, as King (1993: 243) notes in the Sarawak context, 'Dayaks who are employed in the timber industry are unlikely to protest against it'.

Finally, adaptation may be reflected in a decision by grassroots actors to migrate from an area altogether. There are many possible reasons for such action, but there is growing evidence that migration linked to environmental degradation is on the rise, and is fast becoming the major form of adaptation in contemporary times (Jacobson, 1988; Westing, 1992). The decision to move principally for environmental reasons would tend to indicate that the possibilities of adaptation by the means noted above have already been exhausted. Environmental migrants (or 'refugees') often end up moving to urban areas or, in extreme cases, to neighbouring countries, but the common theme is an inability to remain in the home territory due to severe environmental degradation or denied access to needed environmental resources. In Sarawak, for example, some indigenous actors have been forced into abandoning their way of life in the interior of the state for uncertain lives in Malaysia's towns and cities because of the inexorable encroachment of logging operations (Parnwell and King, 1995).

In marked contrast to these diverse forms of adaptation, many grassroots actors have sought to fight the oppression of more powerful actors through what Scott (1985) terms 'everyday forms' of resistance. The following discussion examines the various types of everyday resistance, and evaluates the potential utility and significance of this grassroots response.

Everyday resistance is widely resorted to by poor farmers, shifting cultivators, and the like, when open confrontation with powerful actors carries the real prospect of a massive retaliatory response by the latter. Indeed, the purpose and meaning of everyday resistance becomes clearer when this technique is contrasted with the better-known grassroots or 'peasant' rebellion. Everyday resistance is the antithesis of the peasant rebellion. Whereas peasant rebellion is overt and collective, everyday resistance is covert and often individual; while peasant rebellion directly challenges prevailing political and economic norms, everyday resistance does so indirectly and always on the sly. It is precisely the anonymity of everyday resistance which is, paradoxically, its greatest strength, and yet also its gravest weakness. Everyday resistance may ultimately undermine a detested political and economic order, but it will only do so in the long term, if at all. There are no guarantees, moreover, as to the desirability of the order that takes its place. As a result, the 'weapons of the weak' should not be 'overly romanticised' in that they are 'unlikely to do more than marginally affect the various forms of exploitation [or associated adverse environmental conditions] that peasants confront' (Scott, 1985: 29–30). However, everyday strategies have long been a mainstay of efforts by poor grassroots actors to fight enclosure and environmental degradation throughout the Third World (Colburn, 1989).

Everyday resistance came into its own as a form of grassroots resistance during the colonial era when, as this book has already noted, existing grassroots management regimes were typically disrupted as states enclosed commons resources and businesses, and thereafter exploited those resources as part of a globalising capitalist economy (see above, and also Chapters 3 and 5). In a process that has often been continued to the present day, grassroots actors were denied access to diverse environmental resources and, as political ecologists show, the typical response was one of bitter covert resistance on a day-to-day basis in order to assert local rights to environmental resources in the face of powerful colonial bureaucracies (Guha, 1989; Neumann, 1992; Peluso, 1992; Jarosz, 1993; Bryant, 1997a). Two examples serve to illustrate this argument.

Peluso (1992) describes how the introduction of 'scientific forestry' by the Dutch (using German foresters) in Java in the mid-nineteenth century transformed the lives of those grassroots actors who were dependent on that island's teak forests for their livelihoods. The Dutch colonial state used legal-political and coercive means to wrest control over these prized forests from local people, but in doing so set off a process of resistance that included various 'criminal' acts, notably illegal extraction, general non-cooperation with state officials, and arson attacks on commercial teak plantations and forests. Such resistance was highly varied in practice, with different actors following different strategies depending on the prevailing power relations as well as on local social and environmental conditions. Thus, they

> cut teak despite Dutch efforts to guard the forest. They refused to pay taxes, refused to pay fines, refused to accept wages, refused to leave rented or communal land when their leases expired, refused to participate in the ritual of village reciprocity and the ritual feasts (*slametan*) that accompanied them. Some piled stones in the road they had been ordered to build. The variation in forms of resistance nevertheless expressed a common discontent.
>
> (Peluso, 1992: 71)

A comparable record of everyday resistance occurred in colonial Madagascar where, as Jarosz (1993) shows, French colonial officials sought to stamp out shifting cultivation, but in the process only incurred the implacable opposition of shifting cultivators in this colony. Shifting cultivation (or *tavy*) was a form of long-term land management used for centuries by the Malagasy, but concerns about the possible adverse effects of such cultivation on the island's commercially valuable forests prompted the French to ban this practice in those forests in 1913. As elsewhere in the colonial world (Bryant, 1994a; Jewitt, 1995), this policy was linked to a paternalistic quest to 'civilise' shifting cultivators through a sedentarisation programme that aimed to convert hill-dwelling cultivators into valley-dwelling commercial farmers. However, this 'colonial vision proved difficult to implement' as a result of the widespread resistance of the Malagasy to the restrictions placed on the *tavy* (Jarosz, 1993: 375). Everyday resistance here, as in Dutch-ruled Java, often involved nothing more than the perpetuation of practices that were now illegal, and shifting cultivators were arrested or forced to pay fines for burning and clearing state-protected forests. Indeed, the *tavy* became for the Malagasy a symbol of resistance to French exploitation and control transcending its initial purpose simply as a means of subsistence. A 'culture of resistance' (Peluso, 1992) developed in which the practising of the *tavy* represented a conscious quest to hold on to local culture and beliefs; the fact that such cultivation was undertaken in traditional dress and using traditional tools was a piquant rejection of French attempts to convert the Malagasy to a more 'civilised' European way of life (Jarosz, 1993).

Yet just as techniques of control elaborated under colonial rule were enforced thereafter by postcolonial states (see Chapter 3), so too grassroots actors have continued to resort to everyday resistance techniques as part of a broader effort to protect threatened livelihoods. In many cases, such resistance continues to be associated with 'classic' conflicts over logging, mining or cash-crop production. The postcolonial preoccupation with the construction of large dams is an added activity that has prompted everyday resistance – as well as some of the most bitter public opposition campaigns in recent years (see below). Perhaps the most interesting examples of contemporary everyday resistance, however, relate to efforts by grassroots actors to undermine the practices of states and businesses aiming to 'green' the environment (Lohmann, 1996). This book has noted at various stages the ways in which traditionally powerful actors have sought to retain or enhance their power over other actors through such activities as plantation forestry and eco-tourism. Here, it is useful to highlight how some grassroots actors become disadvantaged through physical displacement or curtailed access to essential environmental resources.

The case of the Maasai, and their link to the killing of 'protected' wildlife in Kenya's game reserves, illustrates this point. The history of the Maasai's exclusion from areas designated as game reserves or national parks is rooted in colonial times when the British severely restricted their ability to pursue traditional semi-nomadic pastoralist practices (Collett, 1987; Peluso, 1993b). In some cases, land was taken from the Maasai by the state and given to large commercial farming interests closely allied to the British, but it was the conversion of vast areas of Maasai lands into national parks or game reserves that has been at the centre of Maasai resistance in recent times. Colonial and postcolonial governments in Kenya have shared the view that Maasai practices are incompatible with wildlife protection. The Maasai practice of grazing cattle on the rangelands, in particular, has been seen by officials as a major threat to endangered wildlife. This perception has never been substantiated with convincing empirical evidence. A lack of evidence, however, did not stop colonial or postcolonial officials from denying to the

Maasai rights of access to parks and reserves. The overriding fear of these officials has been that the 'big game' responsible for rapidly rising tourist numbers might be harmed in some way by the Maasai presence, thereby jeopardising a significant money-earner for the Kenyan state and allied business interests. However, the Maasai have fought this official campaign to marginalise them using various means, notably everyday resistance techniques. Denied access to the watering holes and swamp areas that they and their cattle relied upon, the Maasai began to kill protected wildlife surreptitiously as a means of attacking state and business interests antithetical to their own (Collett, 1987; Peluso, 1993b). According to Peluso, by the late 1970s and early 1980s,

> the restriction of their principal means of livelihood was probably a major reason that some Maasai began killing rhinoceros and elephants in protest. A decade later, some allegedly began collaborating with ivory poachers. They also resisted further appropriation of their access rights by increasing their use of the area surrounding the livestock-free zone, and later demanded tenure rights to all these lands.
>
> (Peluso, 1993b: 205)

The pursuit of an ostensibly 'green' activity by state and business interests (but also some First World ENGOs, see Chapter 6) has thus prompted everyday resistance from adversely affected Maasai pastoralists in a situation subsequently compounded by the arrival of professional poachers involved in the illegal ivory trade.

It remains to consider whether everyday resistance addresses effectively the livelihood concerns of poor and marginal grassroots actors like the Maasai. Scott (1985) certainly warns against exaggerating the likely political impact of everyday resistance or the ability of this strategy to end the dominance of powerful actors (see also J. Scott, 1990). Yet Scott (1985: 35–6) insists that individual acts of everyday resistance by grassroots actors may be 'invisible' to powerful actors, and accordingly seem insignificant, but 'multiplied many thousand fold, such petty acts of resistance by peasants may in the end make an utter shambles of the policies dreamed up by their would-be superiors in the capital'. Such resistance may thus be effective precisely because it is hard to pinpoint, and hence to control. The state in particular faces an acute dilemma in dealing with everyday resistance. On the one hand, it is difficult to single out and punish the perpetrators because of the covert and anonymous nature of their 'crimes'. On the other hand, to publicly acknowledge everyday resistance would be for the state to 'lose face' since such a move would be tantamount to an admission that this actor is unable to exert its authority fully and legitimately in the country (Scott, 1985).

The historical record is ambiguous about the utility of everyday resistance strategies. This has to do partly with the very nature of everyday resistance –

it does not draw attention to itself, therefore few records exist of its successes or failures (J. Scott, 1990). Nevertheless, there are cases that show fairly clearly that everyday resistance has seriously threatened, if not undermined altogether, the policies and practices of powerful actors. In the Madagascar example noted above, for instance, the success of the Malagasy resistance against the *tavy* ban subverted the French vision of large-scale commercial forestry and a docile Malagasy work-force in the commercial economy (Jarosz, 1993). In contemporary Ecuador, in contrast, state efforts to exclude grassroots actors from the country's national parks have prompted extensive clandestine forest clearance by these actors. Such resistance has prompted a re-think on the part of the Ecuadorian forestry department such that this agency is now seeking to integrate grassroots actors in the conservation process through a revamped park management policy. Thus, rules on farming and cutting wood in selected parks have been changed to allow local actors to earn a livelihood from these activities while alternative livelihood options are explored by park officials acting in close consultation with these actors (Bailey, 1996).

The growing emphasis today on public grassroots organisations would nonetheless seem to indicate that everyday resistance and adaptation strategies have clear limits in the quest to resist oppression. Both of these strategies tacitly acknowledge the relative powerlessness and marginality of poor grassroots actors before their more powerful counterparts. In contrast, grassroots organisations often pose a more direct challenge to the latter. Although stopping short of rebellion, they still make plain in no uncertain terms the impatience of poor grassroots actors with the political and economic status quo.

THE POLITICS OF GRASSROOTS ORGANISATIONS

Grassroots organisations are not a new phenomenon. However, it is only since the 1980s that they have become a central means by which grassroots actors fight for social justice and local control over environmental resources. Indeed, grassroots organisations have been in the frontline of the battle for local 'empowerment' (Friedmann, 1992). This section assesses their effectiveness in struggles to change the topography of the Third World's politicised environment. Any attempt to do so must confront the fact that there are hundreds of thousands of grassroots organisations, reflecting an equally diverse range of interests. The following analysis, however, considers the politics of this type of actor by orienting the discussion around two broad categories: protest grassroots organisations and self-help grassroots organisations.

A number of today's grassroots organisations can be traced back to colonial times when anti-imperial protests occasionally took the form of organised public campaigns. In India, for example, the enclosure of communal forests in the late nineteenth and early twentieth centuries by the colonial forest service prompted everyday resistance by grassroots actors. However, it also led to the growth of peaceful non-cooperation movements (forest *satyagraha*) that demanded openly that the British restore local forest rights (Guha, 1989; Shiva, 1991a). These *satyagraha* were usually suppressed by the British, but their affiliation with the urban-based Indian nationalist movement afforded them limited protection in the 1920s and 1930s. Indeed, the latter 'served to legitimize protests oriented towards forest rights, enabling peasants to claim these rights more insistently and with greater militancy' (Guha, 1989: 134). In neighbouring Burma, meanwhile, village nationalist organisations (*wunthanu athin*) organised public protests and civil disobedience against the British in the 1920s and early 1930s. The creation of *wunthanu athin* certainly reflected a broad range of anti-British grievances, but stringent restrictions on the ability of Burmese farmers to obtain forest products from nearby 'reserved' forests were a prominent cause of grassroots disquiet. Here again, grassroots political protest and nationalist agitation proved a potent challenge to the colonial state (Bryant, 1994c).

Certain contemporary grassroots organisations trace their origins back to such colonial precursors – the Chipko movement of northern India being the best known example (see below). Many grassroots organisations have nonetheless been created only recently. This surge in grassroots activity can be explained partly in terms of the intensified social and environmental problems facing poor grassroots actors which have been discussed throughout this book. A further reason relates to the advent of democratic or quasi-democratic political regimes in many countries since the 1980s, and the generally greater ability of grassroots organisations to flourish under such regimes. Authoritarian regimes have certainly (if often unintentionally) prompted the emergence of grassroots organisations in selected Third World countries (Price, 1994; Eccleston and Potter, 1996). It is nonetheless the case that 'the triumph of people power over armed tyrants' (Korten, 1990: 26) in the 1980s marked an important watershed in the political development of these organisations. Poor actors were able – often for the first time – to organise and publicly protest about their plight without the fear of inevitable retribution.

The advent of relatively democratic conditions in the 1980s also enabled self-help grassroots organisations to be established in many Third World countries. The growing ability of these organisations to address local social and environmental problems without prior state support appears to reflect a broader shift in state–civil society relations in the Third World (Peet and Watts, 1996b). As this book has noted in various places, the role of the state as developer and steward of the environment has been subject to increased challenge from actors that include businesses and environmental non-governmental organisations (ENGOs). Many grassroots organisations have added their weight to this challenge in so far as they seek to by-pass the state altogether through the promotion of 'local solutions to local problems'. As Korten notes,

> the 1980s saw a growing rejection of the myth that government is the sole legitimate agent for development decision making and the management of development resources. It is now widely accepted that civil society has an essential, if not central role in both.
>
> (Korten, 1990: 28)

Various factors lie behind the growing political prominence of protest and self-help organisations in the Third World today. Whatever the reasons for their development, however, both types of organisation integrate environment and development concerns, and thus may be readily distinguished from a number of the First World and Third World ENGOs discussed in Chapter 6. That grassroots organisations do so is, of course, unsurprising, given the dependency of most grassroots actors on the environment for their livelihoods (see above).

Of the two, protest grassroots organisations are typically the most visible politically in that their purpose is precisely to publicise those practices of powerful actors considered inimical to the interests of grassroots actors. Such organisations develop as a collective reaction to planned or existing activities that jeopardise the livelihoods of these actors, and connections with national and international advocacy ENGOs are often vital to their success. Yet it would be wrong in most cases to dismiss these protest organisations as single-issue groups akin to the 'not-in-my-backyard' (NIMBY) organisations that have sprung up over many parts of the First World (McCormick, 1995). Rather, these protest organisations seek to integrate environmental and development concerns in such a way as to promote long-term grassroots environmental management. Ideas about 'appropriate' social and environmental changes are usually embedded in their criticisms of the development projects of powerful actors. Two examples of protest grassroots organisations will be used below to illustrate the possible connections between enclosure, environmental degradation, threatened grassroots livelihoods, and the organisation of protest organisations.

The first example relates to a protest organisation that developed in the late 1970s in response to plans by the Brazilian state to build a series of hydro-electric dams on the Uruguai river. As McDonald (1993) notes, these plans required the displacement of over 30,000 rural farmers (a figure later contested by the grassroots organisation as 'too conservative') who derived a livelihood from corn and soya production. However, and perhaps contrary to outside expectations,

> far from opposing the dams, the communities initially looked favourably on the project as important for the region's development
> ... The principal concern of the *atingidos* [the Portuguese word for

those affected by the project] was to receive fair and timely indemnification for all the losses they would incur.

(McDonald, 1993: 86)

Yet as doubts grew about official compensation, as well as about the precise location and scale of the proposed dam project, the *atingidos* in 1979 founded an organisation known as the Regional Commission of People Affected by Dams (CRAB) to protect local interests. This organisation, and notably the news bulletin that it published regularly, 'became an important organising tool through its ability to provide information to rural communities scattered across the affected region' (McDonald, 1993: 87). However, the more CRAB put pressure on Brazil's state-owned energy corporation (ELETROSUL) to clarify grassroots concerns about the dam project, the more autocratic and cavalier that large agency became. The result was an escalating series of confrontations between CRAB and ELETROSUL in the early to mid-1980s as CRAB's position turned from potential support to implacable opposition to the project. In a newly democratic context after 1985, ELETROSUL was forced on to the defensive as CRAB marshalled national and even international support for its cause. In the end, the project was shelved by the Brazilian state: CRAB had become too well organised, and possessed too many supporters, for the state to be able to continue to ignore it (McDonald, 1993). CRAB's victory illustrated in a Brazilian context the political possibilities of 'people power' in a democratising Third World.

The second example of a protest grassroots organisation relates to the Chipko movement of northern India which has become a worldwide symbol of grassroots protest (Guha, 1989; Shiva, 1991a; Gadgil and Guha, 1994). Chipko first developed in the village of Mandal in the Garhwal Himalaya as a result of an incident in April 1973 in which village women, in a desperate bid to protect local ash trees from outside loggers, resorted to hugging the trees to prevent logging operations. Gadgil and Guha note that this initial act of protest

brought to a fore a simmering but widespread resentment among the hill peasantry, directed at state forest polices which had consistently favoured outside commercial interests at the expense of their own subsistence needs for fuel, fodder and timber. Thus the 'Chipko' (Hug the Trees) movement was born. In the following decade, a wave of protests against commercial logging swept the Himalayan foothills, co-ordinated by Gandhian as well as left wing activists.

(Gadgil and Guha, 1994: 104)

As noted, Chipko has strong links to grassroots protests dating from colonial times (Guha, 1989). Yet this broad-based organisation is also quintessentially modern in its skilful use of the media, and of the democratic process in

India, to pursue its case against commercial loggers operating in the Garhwal Himalaya. An ability to tap the support of Indian and First World ENGOs has also been pivotal in this regard. Indeed, it was the combination of traditional and contemporary tactics that helps to explain its political success. The latter was signalled by the Indian state's decision to impose a ban on the commercial felling of green timber in the Himalayan region of Uttar Pradesh following a meeting between Prime Minister Indira Gandhi and the Chipko leader Sundarlal Bahuguna in early 1979, and after several years of regional political protest (Shiva, 1991a). This 'victory' is questioned by Rangan (1996) who suggests that Gandhi used the Chipko campaign for her own purposes – namely, to promote public sector expansion at the expense of the private sector. Thus, in his view, Chipko succeeded because it suited powerful interests within the state, and not because of grassroots pressure whipped up by Chipko's leaders. Further, Rangan (1996: 222) pointedly notes that despite Chipko's activities, the Garhwal Himalaya remains steeped in poverty, and he criticises environmentalists for being 'oblivious to the processes of marginalization continuing in the region'. Chipko has nonetheless become a model for protest grassroots organisations elsewhere in India (and beyond) as these organisations seek to follow a similar path to 'succcess' (Rush, 1991; Gadgil and Guha, 1994; Rangan, 1996).

Self-help organisations provide an interesting contrast to protest organisations. Unlike the latter, the former typically shun the political limelight, and seek to avoid confrontations with more powerful actors through local activities that emphasise 'non-political' issues. The impetus for these self-help organisations is the need to find a collective response to environmental problems that jeopardise the livelihood interests of grassroots actors. The aim is to develop directly sustainable management strategies linked to such things as fuelwood collection, water provision or rubbish removal. Yet, for all their ostensible innocuousness, self-help organisations are deeply political in so far as they constitute a pragmatic challenge to the political and economic status quo, and an implicit critique of those actors that support the status quo. Thus, these organisations have often developed as a direct response to the inability or lack of commitment of the state to alleviate the environmental problems associated with contemporary Third World development. As such, their activities are a continual reminder to all and sundry that the state is not fulfilling its role either as steward of the environment or as the promoter of social justice.

In many Third World cities, for example, self-help organisations have been developed by the urban poor to alleviate some of the social and environmental problems associated with rapid and unplanned urban development. These problems include inadequate water, housing and sanitation facilities, and collectively illustrate how, in the quest to industrialise, states have usually neglected to provide for basic infrastructure or services. The Third World's urban environmental crisis follows directly from this situation, and is reflected in the growth of self-help organisations. Thus, Douglass (1992) notes how in Bangkok one organisation was founded in the low-income Wat Chonglam district, a part of the city where poor households live in informal housing built on stilts over a rubbish-infested river. Local community leaders, assisted by staff at a nearby university, organised an interest-free loan from a local Thai bank in the 1980s to implement community projects designed to improve the local environment, including improved access to the area and rubbish clean-up. This project succeeded in such a way that Bangkok's municipal government declared Wat Chonglam 'community of the year' in 1991 (Douglass, 1992).

Self-help grassroots organisations have also been prominent in rural areas where they tackle diverse issues pertaining to environmental degradation and environmental management (Conroy and Litvinoff, 1988). In many parts of South and South-East Asia, for example, community forestry organisations have been set up in villages in an effort to wrest control of local forests from outside actors as a precondition for long-term locally controlled forest management (Gadgil and Guha, 1994; Colchester, 1994a). In Bangladesh, for instance, forest protection organisations have increased dramatically in number, from 260 in 1986 to 1,944 in 1991, notably as a result of the organisational activities of Proshika, one of the country's leading NGOs. These groups have sought to protect residual forests while simultaneously planting trees suitable for local use. However, they have often become embroiled in conflict with more powerful state and non-state actors whose interests and power are threatened by these changes (Fisher, 1993). In a number of African countries, self-help organisations have sprung up in communities affected by severe drought in order to establish a more efficient water supply system. Livernash (1992) notes, for example, that a village in Senegal started a community development association following the 1973 drought which raised funds to establish a community water system. Following the building of the water system, other projects were undertaken, including the building of a millet mill for women, a pharmacy and a communal horticulture plot.

Self-help grassroots organisations have also developed as a response to restrictive state conservation measures in that they have served as an important means by which grassroots actors, denied access to environmental resources in protected areas, can find alternative livelihoods. In Ecuador, for example, grassroots actors residing in Machalilla National Park traditionally earned a livelihood from charcoal manufacturing, but the establishment of the park in 1979 ended all that. Faced with a series of restrictions on that economic activity imposed by park officials in the name of forest conservation, the community banded together to develop a series of projects relating to horticulture, tourism, and animal husbandry. These projects have sought to devise alternative livelihoods for local residents that will decrease their reliance on the forests (Bailey, 1996).

179

The complex motivations, interests and activities of both self-help and protest grassroots organisations needs to be emphasised. Indeed, some organisations may even combine the traits of protest and self-help organisations. Mexico's *Comite de Defensa y Preseveracion Ecologica* (CDPE) is a case in point. Founded in 1987 to lobby states and businesses and to promote a solution to the chronic pollution of the Tunal river in Durango, the CDPE (and its precursor the *Comite de Defensa Popular* founded in the 1970s) has combined qualities of public protest and pragmatic community work that contributed to that organisation's growing success in the 1990s (Moguel and Velazquez, 1992).

Despite the evident heterogeneity of grassroots organisations, it is nonetheless possible to discern a number of general features about these organisations that merit discussion here as part of an assessment of the overall political significance of this type of actor. The livelihood-driven focus of protest and self-help organisations has already been noted in this chapter. However, other common features relating to the organisation, sources of support and composition of grassroots organisations require further discussion.

To begin with, it is important to emphasise the centrality of organisational ability in the development of grassroots organisations. This point may seem an obvious one. However, while the desire to defend livelihood interests may be a strong impetus for the development of a grassroots organisation, it is rarely sufficient in itself, but rather must be coupled to other factors internal and external to the community itself. As Friedmann (1992: 143) observes, 'the likelihood of a truly spontaneous organization of the poor is very small'. One feature often associated with the development of grassroots organisations is a local tradition of collective action on social and environmental matters. Thus, it was noted above how the Chipko movement may be seen as a descendant of earlier grassroots resistance initiatives in the area. The point here is that contemporary organisations like Chipko are able to build grassroots cooperation quickly in settings where such cooperation is an important part of the local history (Guha, 1989). Such traditions are not necessarily oriented around social protest. They may just as readily encompass self-help networks involving anything from a few families to entire communities in various local 'public works'.

Another feature of many contemporary grassroots groups is a well-organised and democratic structure which can mobilise people and pursue activities quickly, but which is also sufficiently robust to sustain community activities over the medium to long term. All grassroots organisations certainly cannot be deemed 'democratic' since these groups habitually operate in communities that are influenced by unequal power relations (Hirsch, 1995). Further, it is not always the case that grassroots organisations are able to sustain initiatives over anything other than the short term without external support (see below). We would insist nonetheless that a feature common to many grassroots organisations is the ability to reach a local consensus (however difficult that might be in practice) through democratic consultations on which the authority of the organisation is then based. Grassroots organisations thus derive power and authority in part from a sense of moral purpose – by 'protecting grassroots livelihoods' – that seems to be indelibly associated with the entrenchment of democratic principles at the heart of those organisations.

That many grassroots organisations possess democratic structures is not to say that individual leaders or 'opinion-formers' do not often play an important role in the evolution of an organisation's activities. On the contrary, few grassroots organisations lack an identifiable leader or elected committee. Indeed, a number of organisations developed from a nucleus, dominated by a handful of activist leaders at the start, into much larger entities in which management committees have been created, and which are usually accountable to grassroots 'constituents'. Thus, as discussed above, CRAB is a grassroots organisation that has fought dam construction in southern Brazil. Yet its organisational structure changed significantly over the years: 'from a small group of committed leaders in the early 1980s, the [organisation] had expanded its organisational base by promoting local and municipal committees in the affected regions' (McDonald, 1993: 97). An executive council was formed in the mid-1980s to help organise and coordinate the activities of this rapidly expanding organisation, but the council was simultaneously accountable to local representatives elected to a 'grassroots general assembly'.

Reliance on outside actors is a further feature common to most grassroots organisations in the Third World. Indeed, a prominent theme in the work of political ecologists on grassroots organisations is the fact that external actors have often played a key role in their development (Friedmann and Rangan, 1993). The latter are usually middle class and urban-based, and notably include churches, labour unions, universities and ENGOs in both the First and Third Worlds. Outside groups often provide essential technical, financial and organisational support to grassroots organisations, especially (but not necessarily exclusively) in the first few years of their development. In the aforementioned case of grassroots protest over a proposed series of dams in the Uruguai river basin, McDonald (1993: 97) notes that 'in the first years of its existence, CRAB depended quite heavily on help from outsiders...but it eventually developed into an internally strong and democratic organisation'. If anything, the support of outside actors is even more critical for self-help grassroots organisations in the early stages of their development, since these organisations rarely possess the requisite technical or financial capacity to implement development projects. In Ecuador, for example, Agua Blanca's grassroots organisation (noted above) relied heavily on technical and financial assistance from Ecuadorian universities, as well as national and First World ENGOs, when it devised various projects to decrease local dependence on the forests. Indeed, members of the organisation emphasised that the community's horticulture project would have been impossible to implement without the technical support of a foreign ENGO since no person in the village had experience in garden crop cultivation (Bailey, 1996). The argument here reiterates a point made in the previous chapter that many First and Third World ENGOs were developed specifically in order to support grassroots organisations as part of the broader quest for social justice and/or environmental conservation (Chatterjee and Finger, 1994).

The support of ENGOs and other 'outsiders' does not necessarily end with the successful launch of a grassroots organisation. Indeed, it may become even more critical as the latter seeks to 'scale up' its activities, or as powerful actors threaten the survival of the group in question. A case in point is the struggle of the Penan with loggers and state officials in Sarawak for control of the local forests (Hong, 1987; Colchester, 1993; and see above). Protest activities by the Penan since the 1980s have included the sabotage of logging equipment and the construction of roadblocks to prevent the movement of logs from the area. The response of the Malaysian state to these activities has been to repress the dissent, but this move has only prompted the intervention of two Malaysian advocacy ENGOs (Sahabat Alam Malaysia (SAM) and the World Rainforest Movement (WRM)) on the side of the Penan. These ENGOs have sought to put pressure on both the Sarawak and Malaysian states to end logging in Penan areas through national letter-writing campaigns, national and international protests, legal challenges, logistical support for the Penan, and direct lobbying of Malaysia's political leaders (WRM and SAM, 1989).

Yet this example also illustrates the potential limits of grassroots–ENGO alliances. As a result of the intervention of ENGOs, and the attendant 'globalisation' of the Penan struggle, Malaysian Prime Minister Mahathir ordered a crackdown on Penan activists and the NGO sector generally in 1987. That crackdown included press censorship, tough new laws governing civil protests, and the arrest of key leaders as part of a concerted bid to cow the groups involved. Mahathir also defiantly continued to support logging in Penan areas. Once the issue became a matter of not 'losing face' for the prime minister, the likelihood that the Penan would be successful in their struggle faded fast (Eccleston and Potter, 1996).

Finally, we wish to highlight that there is an important gender dimension to the membership of many grassroots organisations in the Third World. As various scholars have noted, an increasing proportion of grassroots organisations are being created and run by women (Sen and Grown, 1987; Sontheimer, 1991; Mies and Shiva, 1993). This phenomenon partly reflects the long-standing exclusion of women from positions of power within the state, ENGOs and even older grassroots organisations (Livernash, 1992). The great vulnerability of poor women to the effects of enclosure and environmental degradation that was noted above is an additional factor explaining women's growing prominence in grassroots organisations. In many cases, women are the guiding force behind self-help organisations as they seek to rectify themselves the social and environmental problems that afflict them and their families. Livernash (1992: 14) notes that 'women are organising to cope with tasks such as providing water and fuel; growing crops; and generating income from food processing, handicrafts and similar activities'. In the village of Saye in Burkina Faso, for example, women have planned and constructed a dam to catch water that falls only during the area's erratic rainy season to reduce time spent fetching water (Livernash, 1992). In rural India, meanwhile, women have been at the forefront in creating protest organisations to protect local livelihoods. The critical role of women in Chipko has already been noted, and was related to the adverse impact on women of logging in the Himalayan foothills. Indeed, 'sheer survival made women support the movement' (Jain, 1991: 165). These examples illustrate that women are often playing a crucial role in the development of self-help and protest organisations in the Third World. This 'feminisation of the grassroots movement' illustrates the important point that grassroots organisations usually develop to assert local control over environmental resources but that, in doing so, they may also aim to represent women's claims for social justice both inside and outside the local community.

The discussion so far has explored the traits of grassroots organisations as well as the political implications of the growing prominence of this type of actor. It remains to consider in the final part of the chapter whether these organisations have succeeded in promoting the local empowerment that is usually at the heart of their programmes.

TOWARDS LOCAL EMPOWERMENT?

Any attempt to evaluate the success of grassroots organisations is fraught with difficulties given the great heterogeneity of this type of actor in terms of interests, size and organisational ability. Yet it is possible to consider at a general level the relationship between grassroots organisations and local empowerment. In doing so, the potential power of these organisations to promote local livelihood interests and environmental conservation needs to be set against a series of factors that may call into question the long-term effectiveness of grassroots organisations.

It would seem clear from the preceding discussion that grassroots organisations possess a number of features that give this type of actor a potential advantage over other actors in addressing the livelihood concerns of the poor. The fact that grassroots organisations are locally organised and run by, and on behalf of, grassroots actors gives them a legitimacy and accountability that is typically lacking in the case of other actors, including many ENGOs. Indeed, while the latter have often been unable to address effectively the livelihood concerns of poor grassroots actors due to the top-down nature of their initiatives, grassroots organisations are often able to implement programmes which the members of these organisations themselves have devised. These organisations also enable grassroots actors to benefit from 'economies of scale' in social activities. Not only do grassroots organisations facilitate collective action in terms of practical self-help projects, but they also more generally 'provide members with greater financial and negotiating leverage than they would have individually' vis-à-vis more powerful actors (Livernash, 1992: 221). In addition, the creation of a grassroots organisation serves as a focal point for the introduction of external financial or technical support from ENGOs and other actors, whereas prior to the creation of such an organisation the provision of external assistance would have been a hit-and-miss affair. Indeed, the role of grassroots organisations as conduits for outside assistance reiterates the importance of organisational ability in the tackling of grassroots livelihood concerns. Finally, grassroots organisations may serve as an effective vehicle whereby individual grassroots actors can voice their political opposition to the practices of traditionally powerful actors without fear of physical or economic retaliation. To be sure, the growing prominence of grassroots organisations has not eliminated the persecution of the weak by the strong everywhere (Broad, 1993). Yet as grassroots organisations become stronger and develop links to actors external to the community, they often gain the ability to turn the tables on local 'bosses' and other agents of oppression (but see below).

These factors suggest that grassroots organisations may be an effective way in which to promote the local empowerment of poor farmers, pastoralists and other grassroots actors. There is certainly evidence of political victories by grassroots organisations in the Third World. Whether it be the anti-dam struggle of CRAB in southern Brazil or the development projects of the people of Agua Blanca in Ecuador, the literature suggests that the ability of self-help and protest organisations to promote local empowerment for their members may be growing (Fisher, 1993). Yet, there are other factors that encourage a more cautious assessment of the ability of these organisations to transform the topography of the Third World's politicised environment.

First of all, there is the problem of the limited scale and impact of many of the projects developed by grassroots organisations. As Livernash (1992: 222) notes, 'local organisations are usually small in scale and may therefore have limited impact and little interest in scaling up their activities or influencing government strategies for poverty alleviation'. This 'Achilles heel of localization' (Esteva and Prakash, 1992) may constrain the effectiveness of grassroots organisation initiatives in that those initiatives do not assist a significant number of needy people. This situation may slow considerably the ability of grassroots organisations to effect change in local political and ecological conditions. To be sure, and as Chapter 6 noted, First and Third World ENGOs may be critical here in helping grassroots organisations to overcome such issues. Yet it was also suggested in Chapter 6 that ENGOs do not necessarily share the livelihood interests of grassroots organisations. More importantly, there are definite, if often ill-defined, constraints on the ability of ENGOs to come to the assistance of grassroots actors in the Third World. As Vivian (1994) observes, there are 'no magic bullets' where the abilities of ENGOs are concerned.

Second, even with the assistance of ENGOs, there are no guarantees that grassroots organisations will be able to overcome the serious political and economic obstacles that traditionally powerful actors may throw in their way. At the local scale, political 'bosses' often seek to subvert the autonomy of grassroots organisations that they usually regard as a major threat to their political or economic position. In the Philippines, for example, there are thousands of grassroots (or 'peoples'') organisations, but many of them have found it exceedingly difficult to dislodge local bosses involved in economic activities that degrade the environment (Broad, 1993; Vitug, 1993). Further, and notwithstanding the official support that many states and multilateral institutions are now according to grassroots organisations, these actors persist in the promotion of activities such as plantation forestry or eco-tourism that often result in the continued marginalisation of poor grassroots actors (see Chapters 3 and 4). Growing activity by grassroots organisations thus does not in itself signal a shift in power relations between the weak and the strong in many Third World countries. The persistence of logging in Sarawak or the continued (albeit modified) development of dams along the Narmada river in India bear eloquent testimony to the persistence of unequal power relations that grassroots campaigns have scarcely begun to alter (Gadgil and Guha, 1994; Eccleston and Potter, 1996).

Third, grassroots organisations are not always unified in their aims, interests or even philosophies. Hence, internal divisions may weaken the political effectiveness of these organisations vis-à-vis other actors. In so far as grassroots organisations base their operations on democratic principles, they must confront the heterogeneity of grassroots interests, and the very real prospect that those interests may not be compatible. In the case of Chipko, for example, conflict within that organisation developed in one village when the men sought to fell a community oak forest in order to establish a potato-seed farm from which they would primarily benefit, while the women voraciously opposed this move on the grounds that they would thereby lose their main local source of fuelwood and fodder (Agarwal, 1992). Further, as grassroots organisations develop self-help projects they must confront the possibility that some members may seek to benefit from these projects without putting in a 'fair' share of the required effort – the classic 'free-rider' problem. In Agua Blanca's grassroots organisation in Ecuador, for example, one conservation-development project failed shortly after its introduction primarily because members could not agree among themselves about the amount of time that each member should allocate to the project (Bailey, 1996). The long history of CPRs in many parts of the Third World certainly illustrates that conflict within grassroots organisations can be overcome. However, as projects get under way, and begin to succeed, the prospect of internal dissent within these organisations is an ever-present threat – especially in a context in which powerful local enemies of these organisations (i.e. bosses) may seek to subvert their activities.

Finally, the dependency of most grassroots organisations on the support of ENGOs, states or multilateral institutions suggests that the activities of the former are extremely vulnerable to disruption in light of the shifting interests of the latter. This chapter noted that the success of grassroots organisations has been linked to the provision of financial and technical support from assorted benefactors, especially ENGOs (see also Chapter 6). Yet what is given today may be withdrawn tomorrow, thereby introducing uncertainty as an inherent feature of the activities of most grassroots organisations. Further, reliance on the support of outsiders has meant that sometimes the decisions made regarding the design and implementation of projects may come from the outside support agency, rather than from within the grassroots organisation itself. In these cases, what may have been intended originally to be a process whereby grassroots actors are empowered turns out in practice to be merely participation by organised grassroots actors in a programme planned and executed by outsiders (Vivian, 1994). External dependency may translate all too readily into a situation in which grassroots organisations are manipulated by other actors to suit the interests of the latter rather than serving as a temporary phase in which the grassroots organisations are allowed to develop autonomous objectives and capabilities.

The political significance and effectiveness of grassroots organisations is thus far from clear. They may represent an effective means to guarantee that the livelihood interests of grassroots actors receive social priority. Then again, they may be a largely ineffectual enterprise whose activities are mainly defined by outside actors' perceptions and interests. However, if political ecologists highlight the ambiguities and contradictions associated with the quest for local empowerment by grassroots organisations, they have also shown that grassroots actors now have an increasing array of means to challenge the policies and practices of powerful actors (Ghai and Vivian, 1992; Friedmann and Rangan, 1993). In addition to the possibilities afforded by adaptation and everyday resistance, grassroots actors can also pursue their interests through public fora, notably through grassroots organisations. That the public struggles of grassroots actors are receiving considerable media attention also suggests that the message propounded by grassroots organisations – that environment and development concerns be linked in a manner beneficial to the poor – is being more widely heard.

As a result, traditionally powerful actors have been finding it necessary to take ever greater notice of the needs and interests of grassroots actors. Indeed, states and multilateral institutions are under growing pressure to re-orient their policies so as to 'put the last first' (Chambers, 1983). Even businesses are increasingly being forced into justifying how their activities are likely to affect the poor. Our discussion of these various actors has nonetheless suggested a need to avoid hasty assessments about possible changes to the unequal power relations that have benefited mainly states, businesses and multilateral institutions. Nothing said in this chapter alters that view. If the topography of the Third World's politicised environment is changing, it is doing so at best slowly and in ways that probably belie easy description.

Political ecologists will need to be increasingly sensitive to the complexities associated with such change. We would argue that they must also broaden their focus so as to conceive of the Third World's politicised environment in its totality, rather than in the selective and mainly 'land-based' manner that has been the case so far. It is to the possible contours of a more inclusive Third World political ecology that we now turn.

CONCLUSION

This book has sought to understand the politics of Third World environmental change from the perspective of political ecology. In the process, the aim has also been to assess the contribution of that emerging research field to an appreciation of the causal forces that lie behind the Third World's mounting environmental crisis. Our central argument has been that the most useful way to go about achieving both objectives is to think in terms of the role of various actors in relation to a politicised environment characterised by unequal power relations.

The decision to use an actor-oriented approach in this introduction to Third World political ecology was motivated chiefly by a concern to 'put politics first' – and our associated belief that an understanding of the political interests and actions of key types of actors is the very stuff of politics. In adopting this approach, we do not intend to imply that other political-ecological approaches based on the analysis of environmental problems, concepts, geographical regions or socio-economic characteristics are not valid. Rather, these different approaches need to be seen as complementary to each other, and to the actor-oriented approach used in this book.

Nonetheless, the latter has usefully highlighted several broad themes that do not necessarily emerge when alternative political ecology approaches are adopted. First, this book has highlighted the essential complexity of the interests of actors as they interact with each other and with the environment in the Third World. Thus, for example, Chapter 3 describes the diverse political, economic and strategic considerations that have influenced the role of the state as a major contributor to environmental degradation. The state has been an important actor in promoting capital accumulation, as political ecologists have long suggested, but equally, if not more importantly, it has also been an actor with its own political and strategic interests that have not always been congruent with those of capital. In contrast, Chapter 5 illustrates that, while businesses share a common economic interest in the maximisation of profit and market share, other political and social interests, linked notably to winning the confidence of state leaders or consumers, feature increasingly prominently in the thinking of business executives. The point of the discussion in this chapter is not to suggest that businesses attach a low priority to economic interests – far from it – but rather to suggest that the interests of businesses cannot be reduced in a simplistic fashion to economic factors.

Second, this book has shown that, just as conflict between actors reflects divergent interests, so too conflict exists between individuals or groups of individuals within each category of actor, conflict which is based on differing interests and concerns. In the case of the state, as Chapter 3 suggests, institutional conflict has been associated with the internal organisation of states, as well as with their territorial definition. Thus, the worldwide triumph of the functionally defined state has been associated with the creation of entrenched, and at times contradictory, bureaucratic interests and attendant intra-state conflict. The territorial definition of the state, meanwhile, has involved the evolution of a 'community of states' in name only as the pursuit of national interests has resulted in both inter-state conflict and intensifying environmental degradation at all scales. Yet the state is not the only actor subject to internal strife. Chapter 6, for example, illustrates that the environmental non-governmental organisation (ENGO) community is similarly riven by divergent interests. Differences within the ENGO community notably centre on the question of finance, as individual ENGOs compete with each other to survive in a context of limited funds. However, the discussion of the differences between First and Third World ENGOs also notes conflict between these two groups over who is to control the content and direction of ENGO campaigns in the Third World.

Third, this book has suggested that there are decided, if not always particularly clear, political implications associated with the specific organisational traits of the different actors involved in Third World environmental change and conflict. This point is most evident with regard to the state, an actor which derives a good deal of its power from its role as the organisation that is responsible for promoting economic development and environmental conservation within a defined territory, and which is granted a 'formal monopoly on the means of coercion' in order to do so. That role is being increasingly challenged by other actors to be sure, but challenges to its authority do not alter the fact that the state is the only actor that plays that legal-political role, and thus potentially benefits from the sources of political power that flow from that role (Mann, 1986; Hobsbawm, 1996). In contrast, grassroots actors derive their power primarily from the combination of a detailed local social and environmental knowledge, and a willingness and determination to use such knowledge through covert and public means to promote their interests. These actors have a variety of means at their disposal to resist more powerful actors; this ensures that, however ostensibly weak they appear, they are nonetheless rarely, if ever completely, without power (Scott, 1985). Yet if grassroots actors derive their power largely from the 'place-based' nature of their activities and interests, this trait can also be a significant political weakness, even an 'Achilles heel of localization' (Esteva and Prakash, 1992) as is noted at various stages in this book.

Finally, this book has suggested that there is often an inherent logic to the coalitions and alliances that develop between different actors based on their contrasting traits and interests. Thus, for example, a recurring theme in this book, and in the political ecology literature generally, is the 'natural' alliance between states and businesses on the one hand, and between grassroots actors and ENGOs on the other. While one of our goals has been to question the 'naturalness' of these coalitions by highlighting the tensions and conflicts between actors, we have nonetheless presented evidence that tends to suggest that cooperation between the actors in each of these groups is especially likely due to complementarities in their interests and traits. Thus, states use their legal-political powers to grant businesses privileged access to environmental resources, while businesses use their financial and technical knowledge 'efficiently' to extract, produce and market environmental resources and/or consumer goods. Both actors seek to expand commercial activity so as to increase their income and/or power over other actors. In contrast, grassroots actors use their grasp of local political-ecological conditions to resist more powerful actors, whereas ENGOs seek to provide technical and financial support, as well as media coverage, to these location-specific struggles. Both grassroots and ENGO actors here seek to assert the primacy of community environmental management so as to promote social justice and/or environmental conservation.

It needs to be emphasised that these themes will be more or less apparent depending on diverse location-specific political, economic or ecological factors. They serve nonetheless as a useful general reference point in the quest to understand the causal forces at work in the Third World's environmental crisis. That crisis, in turn, needs to be related to an appreciation of the essentially politicised nature of the environment in the Third World.

We have suggested in this book that the work of political ecologists, whatever approach that they adopt in their research, is largely about seeking to explain the topography of a politicised environment. To think about the environment as a 'politicised environment' helps to overcome the human–environment dichotomy that is a major weakness in many other environmental research fields (Szerszynski et al., 1996). It serves as a reminder of the essential integration of human activities and environmental processes – a central theme in geography in general and in political ecology in particular. As Chapter 1 suggests, it is this integrated approach that also helps to distinguish 'political ecology' from 'environmental politics'. It is true that a common interest in politics and the environment leads to a certain amount of overlap between the two fields, especially surrounding the role of the state in environmental change and conflict. However, while

CONCLUSION

environmental politics tends to focus almost exclusively on analysing that role, political ecology considers the much wider interactions of state and non-state actors with each other (including the interactions between non-state actors) and with the physical environment. If 'radical' scholars within political science are probing the analytical possibilities of 'de-centring the state' and 'civic politics' (e.g. Walker, 1993; Wapner, 1996), this important work has yet to alter the essential 'state-centrism' of that discipline in general, and of the sub-field of environmental politics in particular (e.g. Young, 1992; Garner, 1996).

The idea of a politicised environment is thus central to the distinctive contribution of political ecology to an understanding of the politics of environmental change in the Third World. The full implications of that idea nonetheless bear reiteration. It is, above all, a specific acknowledgement of the growing human production of 'nature', and the political forces behind such production. If political ecology is an inquiry into 'the political sources, conditions and ramifications of environmental change' (Bryant, 1992: 13), then the role of power in the mediation of relations between actors over environmental matters becomes of paramount consideration. Such power is reflected variously in the ability of an actor to control access to environmental resources, shift the human exposure to hazards primarily on to other actors, direct societal resources (via the state) into certain projects, but not into others, and through control of the 'public transcript' seek to regulate the discursive representations of environmental change. Power relations are thus inscribed in the environment, and in environmental ideas, but however unequal those relations are, power is always a two-way process (Giddens, 1979; J. Scott, 1990; Peet and Watts, 1996a). Indeed, a central concern of this book in adopting an actor-oriented approach has been precisely to illustrate the complexities of power relations between actors over environmental issues in the Third World.

The distinctive contribution of Third World political ecology is thus to emphasise the role of politics in Third World environmental change – indeed, to conceive of such change as an inherently political process. Thus, as scholars from diverse disciplinary and ideological backgrounds speak increasingly of the need to 'reinvent' our understanding of nature in terms of seeing nature as an essentially human construct (Cronon, 1995; Castree, 1995), political ecologists are well placed to assert the primacy of politics in such a revised understanding. As this book has shown, political ecologists have already drawn upon a diverse theoretical literature in order to elaborate that revised understanding. Beginning with neo-Marxism in the late 1970s and early 1980s, they moved on in the late 1980s and 1990s to incorporate thinking derived from neo-Weberianism, feminist and new social movements theorising, as well as (latterly) poststructuralism and discourse theory. Political ecologists have stretched the meaning of 'political economy' in the process so as to take into account a series of important themes relating

191

notably to state autonomy, gender conflict, everyday resistance and discursive formations.

Concern over the 'theoretical untidyness' of political ecology has nonetheless prompted Peet and Watts (1996b) to urge that the field turn collectively to discourse theory – or what Escobar (1996) terms 'poststructural political ecology' – for intellectual clarity. Peet and Watts suggest that

> one of the great merits of the turn to discourse, broadly understood, within political ecology, is the demands it makes for nuanced, richly textured empirical work (a sort of political-ecological thick description) which matches the nuanced beliefs and practices of the world.
>
> (Peet and Watts, 1996b: 38)

The important role of discourse in conditioning political-ecological conflicts is not to be denied. Indeed, it has been a recurring theme in this book as we have sought to relate how conflict over environmental resources is also typically a struggle over ideas as to what constitutes 'appropriate' environmental use and management. We are nonetheless concerned that a 'turn to discourse' may result in a turn *away* from the material issues that, after all, prompted the birth of Third World political ecology in the first place (we would also question the utility of yet another 'ecology' – this time 'liberation ecology' – when a 'robust' political ecology will suffice, see Watts and Peet, 1996). It is important to appreciate discursive formations precisely because they have something potentially very interesting to say about the material practices of actors involved in social (and environmental) conflicts – a point not always clearly understood in a discourse literature prone to abstractness, if not abstruseness (e.g. Bhabha, 1994). The theoretical argument of Peet and Watts (1996b) will undoubtedly generate much (welcome) debate among political ecologists. However, it is unlikely that it will win many adherents without a more rigorous articulation of central theoretical principles, especially surrounding the relationship between discourse and practice.

While there is a need to engage in theoretical debate over the principles that might guide political ecology into the next century, we would suggest that it is also crucial that further thought be given to empirical questions, notably concerning the range of subjects that political ecologists are prepared to study. The next section thus calls for a more inclusive subject-matter for Third World political ecology.

AN EVOLVING RESEARCH FIELD

We suggested at the start of this book that Third World political ecology will need to expand beyond its anthropological-style focus on land management problems if it is to remain relevant to the changing dynamic of political-ecology conflicts in the Third World. However, our argument is not so much a call for political ecologists to abandon the analysis of such problems, but rather a plea that they also tackle non-land-based problems and conflicts more systematically as part of a comprehensive treatment of the political ecology of the Third World's environmental crisis. That crisis is not synonymous with a land crisis, but rather encompasses a diversity of specific environmental problems that manifest themselves in terms of alterations in land, water or air quality – alterations, moreover, which separately and collectively have an unequal impact on humankind. Political ecologists ought to be at the forefront in analysing all of these problems. In what follows we sketch very briefly some of the issues germane to an evolving research field that would seek to heed this plea.

Political ecologists have been especially good at addressing issues linked to alterations in land quality. For example, they have explored extensively the political and ecological implications of deforestation (and now increasingly reforestation), rangeland degradation and soil erosion (Blaikie, 1985; Hecht and Cockburn, 1989; Peters, 1994; Lohmann, 1996). However, they have been relatively negligent in examining comparable questions arising from alterations to the quality of water, and especially air. To be sure, a few studies have considered the question of water quality and availability, but usually only as an adjunct to the primordial issues of land management and conflict (e.g. Sheridan, 1988). In contrast, very little systematic work has been conducted on the political ecology of water use and management per se. Much of the work that has been conducted on this topic has concentrated on inter-state 'hydropolitics' in a Middle Eastern and North African context (e.g. Waterbury, 1979; Porter and Brown, 1991), yet this leaves a vast subject-matter relating to the role of water in human affairs, and how control over water use is linked to unequal power relations, largely unexplored. How do powerful actors seek to manipulate both the supply and quality of water in society to promote their own interests at the expense of other actors in a context frequently characterised by water scarcity? The issues here are not new, but they nonetheless await a full investigation by political ecologists – although work by Swyngedouw (1995, 1996) in an Ecuadorian context provides fascinating preliminary evidence of the utility of research on this topic.

Changes in air quality represent another largely neglected research topic. A growing literature certainly addresses the issue of urban air pollution as part of a broader attempt to assess the Third World's growing urban environmental crisis (e.g. Hardoy et al., 1992; Setchell, 1995). Yet much of this literature is largely descriptive, and oriented towards an urban planning readership. Accordingly, there is little appreciation of the specifically political dimension to this crisis. Yet unequal power relations are as likely to be 'inscribed' in the air as they are to be 'embedded' in land or water, and the seemingly ephemeral nature of air 'degradation' should not blind us to its often enduring political impact. That impact may be reflected in terms of a First/Third World divide with selected Third World countries serving as 'pollution havens', but is also linked to the prolonged exposure of poor grassroots actors to air pollution – and associated medical illnesses (Hardoy *et al.*, 1992). However, political ecologists have scarcely begun to explore this particular 'politicised environment'.

As political ecologists move to incorporate the investigation of air and water degradation issues in their work, they will almost inevitably need to readjust their analytical focus in other respects. Thus, greater attention to these issues will serve to emphasise the relatively neglected topic of urban political ecology. In a context of rapid Third World urbanisation, it is startling to realise how few political ecologists have addressed this topic. To be sure, there is a growing literature, oriented notably around the journal *Environment and Urbanisation*, that addresses air, water or land degradation and management in an urban context. Yet, as noted above, much of this work does not adopt a political ecology perspective, and thus fails to address adequately questions of unequal power relations in the context of the urban environment (Douglass, 1992 and Swyngedouw, 1995, 1996 are notable exceptions). There is thus a need to correct the long-standing 'rural bias' in Third World political ecology through research on urban environmental change and conflict. However, the interconnections between urban and rural issues must not be forgotten in the process since the interpenetration of urban and rural areas appears to be a crucial, but often overlooked, trend in Third World development (see Rigg, 1997 for a fascinating South-East Asian exploration). One way in which to explore the urban–rural nexus might be to develop the idea of a 'political-ecological footprint' – the 'ecological footprint' idea propounded by William Rees and others, but with a 'political twist' (Rees and Wackernagel, 1994).

Greater attention to air and water problems and conflicts will also encourage a more frequent application of the political ecology perspective at the regional and global scale. A feature of research in Third World political ecology to date has been that an emphasis on land degradation questions has encouraged political ecologists to focus (naturally enough) on the local scale. As this book has shown, they have certainly tried to link these essentially local research agenda to the wider political and economic context. Yet, however effective that link may be, such work still does not address those aspects of the politicised environment that can only be fully appreciated as environmental problems operating simultaneously at the local, regional and global scales (see Chapter 2). The problem of air pollution, for example, is associated with location-specific conflict among actors linked to the production of this problem. However, it is also a growing regional and even global problem that equally merits analysis from a political-ecology perspective. There are 'winners and losers' associated with global climate change, but it is likely that the Third World 'will be hit harder' than the

First World (Meyer-Abich, 1993: 78) – thereby reiterating the utility of a specifically 'Third World' political ecology perspective on this and other global issues (see Introduction). In an era characterised by the 'globalisation' of production relations, as well as a number of associated environmental problems, the need could not be greater for a perspective able to assess the implications of these globalising processes for the Third World.

In the process, political ecologists will need to address questions raised by Beck (1992) about the development of a risk society linked in part to unseen environmental changes prompted by industrial activity. What we termed in Chapter 2 the systemic dimension to a politicised environment has received very little attention in Third World political ecology as scholars have focused for the most part on the everyday and episodic dimensions of that environment. Yet the argument that everyone around the world is exposed potentially equally to, say, hazardous nuclear emissions or dangerous concentrations of pesticides associated with 'industrial society' merits far greater attention from political ecologists operating in a Third World context than has hitherto been the case. How does this argument square, for example, with the view of the Third World as playing the role of a pollution haven in a 'global industrial society'? How much weight is to be attached to claims of a generalised risk when, in a broader Third World context, a highly unequal distribution of environmental risks has long been the order of the day? Further, what, if any, are the political implications of generalised risk in the Third World? Political ecologists bringing to bear insights derived from research in the Third World may have interesting things to say in general about the 'risk society'.

A POLITICALLY ENGAGED POLITICAL ECOLOGY?

It is worth concluding this book with a few final comments about the question of the practical role and purpose of political-ecology research in the Third World. Who is the intended audience for this research, and what do political ecologists hope to achieve through their research?

It is our firm belief that Third World political ecology is a research field that seeks to explain the topography of a politicised environment, and the role that diverse actors play in the 'moulding' of that environment, so as to better assist those actors in society who are fighting for social justice and environmental conservation. As we noted in the Introduction, political ecologists tend to be rather reticent about the possible contours of an 'ideal society'. Political ecologists tend for the most part to describe problems rather than prescribe solutions. It may well be that this role is the most important contribution that scholars who adopt a Third World political-ecology perspective can possibly make, on the principle that in order to resolve the Third World's environmental crisis one must clearly first understand the nature and dynamic of that crisis.

The case for a more politically engaged Third World political ecology is nonetheless compelling. There is, of course, a large literature on 'action-oriented' and 'participatory' research that is designed to 'put the last first' (Chambers, 1983). The suggestion here is not that political ecologists are unaware of this literature and its practical implications, but rather that the time is now ripe for the research field as a whole to integrate those implications systematically into its remit, and for future research to reflect increasingly the practicalities of political engagement. Given the political implications of research in the field, for example, what 'policy' advice should political ecologists give, and more importantly, to which actors should such advice be proffered?

There are no easy answers to this question. Thus, it would be tempting for political ecologists to target their work, and any practical political efforts that flow from that work, exclusively at poor grassroots actors and ENGOs. This approach would agreeably stand in sharp contrast to the approach of many mainstream scholars who typically conclude their work with appended policy prescriptions to the state or multilateral institutions – that is, to those traditionally powerful actors that, as political ecologists often suggest, are at the root of the Third World's environmental crisis. Yet, as this book has attempted to show, such an approach would be misguided in that it would overly simplify a highly complex situation. The main argument of this book is that the Third World's environmental crisis needs to be understood as an outcome of the interaction of actors operating in a context of unequal power relations. Those actors, in turn, are motivated by often quite complex interests and objectives. As such, a more appropriate course would appear to be political engagement on several fronts, at different scales and with various actors, as part of a multi-faceted campaign on behalf of environmental conservation and social justice.

In the end, however, it would be wise not to exaggerate the potential impact of such intervention by political ecologists. The attainment of social justice and environmental conservation will ultimately be achieved by a vast array of actors operating at the local, regional and global scales, and among this great assemblage, the activist scholar will be indeed a relatively small figure. Yet, a recognition of this fact should not be a counsel for despair (or a call for a career change!), but rather a humbling reminder of the limitations of all scholarship in the face of rapidly changing political and environmental conditions in the Third World. It is with this sentiment in mind that we hope that this book will be of use not only to those scholars who are concerned about the Third World's environmental crisis, but also to all those actors whose efforts will ultimately contribute to the resolution of that crisis.

A GUIDE TO FURTHER READING

The following discussion provides a selective guide to key works in Third World political ecology as well as sources of theoretical pertinence to research in the field. It is intended to provide a concise introduction to the literature mainly for those new to the subject-matter. Full references are located in the Bibliography.

THEORETICAL WORKS

There are a growing number of books and articles that explore theoretical issues surrounding the 'political economy of the environment'. Johnston (1989) and Harvey (1993) provide useful overviews of the links between capitalism, the state and environmental degradation. The environmental 'contradictions' of the capitalist system are explored further in the work of J. O'Connor (1988) and M. O'Connor (1994a), as well as by contributors to the journal *Capitalism, Nature, Socialism*. Few attempts have been made so far to specify what an 'alternative political economy' would look like. Two recent notable exceptions are Pepper (1993) and Norgaard (1994). Beck (1992) explores generally the notion of a 'risk society' (i.e. 'systemic' dimensions) that has potentially important implications in a Third World context. A lively mainstream critique of 'radical scholarship' is to be found in Lewis (1992).

THIRD WORLD POLITICAL ECOLOGY

General works

There are relatively few 'general' books or articles in this research field since most political ecologists have eschewed generality in order to focus on specific empirical questions. Several authors nonetheless explore aspects of the 'politicised environment' in general or conceptual terms. Key works by Blaikie (1985) and Blaikie and Brookfield (1987) locate the issues of soil erosion and land degradation (i.e. 'everyday' dimensions) in a broad political-ecology context, while Blaikie *et al.* (1994) elaborate the connections

197

between 'natural' hazards and people's vulnerability (i.e. 'episodic' dimensions). The link between ideas, discourses and the political economy of environmental change is an important theme. Thus, Redclift (1987), Adams (1990) and Escobar (1996) explore the ambiguous mainstream concept of 'sustainable development', while Schmink and Wood (1987), Blaikie (1995b), and Peet and Watts (1993, 1996b) consider the broader ambiguities of environmental ideology, discourse and practice. Debates in Third World political ecology pursue diverse trajectories, but include questions as to the likely role of 'poststructuralism' (Escobar, 1996; Peet and Watts, 1996b) and the 'new ecology' (Zimmerer, 1994) in the field's development. Bryant (1992) provides a review of the literature emphasising key themes (contextual sources, conflicts over access and political ramifications) in the research. Finally, several empirical volumes merit attention here for their general and comparative insights. These include Repetto and Gillis (1988) on state complicity in deforestation, Little and Horowitz (1987), and Friedmann and Rangan (1993) on local-level conflicts, and Neumann and Schroeder (1995) on the contradictions between local rights and global environmental agenda. Watts and Peet (1993), subsequently revised as Peet and Watts (1996a), provides an eclectic, but stimulating collection of theoretically informed studies. Finally, the *Ecologist* magazine remains an indispensable aid to research on political-ecological conflicts. There is no major English-language academic journal dedicated explicitly to the subject as yet, but political ecologists publish their work regularly in the following geography and development journals: *Political Geography, Economic Geography, Antipode, Geographical Journal, Society and Space, Annals of the Association of American Geographers, Development and Change, World Development, Society and Natural Resources*, and *Capitalism, Nature, Socialism*.

Specific works

The following entries are a selective list of those works especially pertinent to specific actors. It needs to be noted nonetheless that many empirical studies explore the interaction of several actors in a given locality – and hence, may have applicability under various headings.

The State

Political ecologists have explored in considerable detail the important role of the state in Third World environmental change and conflict. Repetto and Gillis (1988), and Hurst (1990) offer useful overviews of the range of incentives that this actor has provided for logging and other 'development' purposes. The 'strong states' of selected Latin American and Asian countries have been the subject of particular attention. Thus, Bunker (1985), Hecht

(1985), Hall (1989), Hecht and Cockburn (1989), and Schmink and Wood (1992) provide insightful accounts of the ubiquitous social conflict and environmental degradation generated by state-sponsored development projects in the Amazon region of Brazil. In a South and South-East Asian context, meanwhile, Guha (1989), Peluso (1992) and Bryant (1997a) examine state involvement in the forest politics of India, Java (Indonesia) and Burma, respectively. An important paper by Peluso (1993b) links Asian and African case studies to evaluate the 'coercive conservation' practised by many states in the Third World. Such coercion is frequently linked to policies and practices first introduced by colonial states. The Asian case studies noted above explore this latter theme, as does Watts (1983a) in a landmark case study from northern Nigeria. Other scholars also consider colonial interventions notably in Madagascar (Jarosz, 1993) and Tanzania (Neumann, 1992), while contemporary state actions are covered by Little and Horowitz (1987), Bassett (1988), Moore (1993) and Neumann (1995), among others. Finally, Walker (1989), Johnston (1989) and Hurrell (1994) provide contrasting general insights about state environmental management.

Multilateral institutions

The growing influence of multilateral institutions on human–environmental interaction in the Third World is now widely remarked on in the literature. Much work has concentrated on the international financial institutions (IFIs). A recurring theme is thus the link between the lending policies of these institutions and Third World environmental degradation. Rich (1994), and George and Sabelli (1994) critically evaluate the activities of the World Bank – the 'largest development agency in the world', while Adams (1991), George (1992), Reed (1992) and Kendie (1995) consider the involvement of various IFIs in the debt–structural adjustment–environmental degradation nexus. However, political ecologists have paid little attention so far to 'technical' multilateral institutions, and the associated 'political economy of knowledge transferral'. Marshall (1991) provides an initial assessment of the role and impact of the Food and Agriculture Organisation in the Third World (see other articles in the same issue of the *Ecologist*). Multilateral institutions operate in an international context in which global treatment of environment and development issues is often shaped by inter-state conferences and summits. The subject of 'environmental summitry' is addressed by such authors as Hecht and Cockburn (1992), Sachs (1993), Middleton *et al.* (1993), and Chatterjee and Finger (1994).

Business

The role of business in shaping Third World environmental change and conflict has yet to receive the attention that it deserves in Third World political ecology. Scholars certainly emphasise state–business alliances in their work, but rarely elaborate the history, organisational structures or policies and practices of either transnational corporations (TNCs) or local Third World businesses. The essays in Pearson (1987) nonetheless provide an overview (especially Gladwin, 1987), while Korten (1995) examines TNCs in detail. Various authors provide brief snapshots of big-business practices – for example, Schmink (1988) and Hurst (1990) in the Brazilian and South-East Asian contexts, respectively. Marchak (1995) and Lohmann (1996) provide more detailed accounts of the pulp and paper industry in Thailand, Indonesia, Brazil and Chile. Research featured in the *Ecologist* regularly 'puts the spotlight' on major TNCs. For example, Moody (1996) investigates the mining firm RTZ, while Kneen (1995) assesses the impact of the cereals giant Cargill. Eden (1994) addresses cooperation between TNCs designed to strengthen their global bargaining power vis-à-vis other actors. The recent phenomenon of Third World-based TNCs operating elsewhere in the Third World is considered in the Guyanese context by Colchester (1994b).

Environmental non-governmental organisations

A number of political ecologists document the impact of this relative newcomer to Third World environmental change and conflict. Environmental non-governmental organisations (ENGOs) are often lumped together with 'grassroots organisations' in a manner that belies their distinctive origins and practices. Korten (1992), Clark (1991), Carroll (1992), Friedmann (1992), and Peet and Watts (1993, 1996b) offer useful introductory perspectives on the 'NGO movement'. Fisher (1993) provides an important comparative account of ENGOs in a study rich in empirical examples from throughout the Third World. Other writers focus on specific ENGOs or regional contexts. Thus, Wapner (1996) provides helpful case studies of Greenpeace, the World Wildlife Fund and Friends of the Earth, while Rush (1991) and Price (1994) review ENGO activity in South-East Asia and Latin America, respectively. Permpongsacharoen (1992), meanwhile, provides an 'insider's account' of Project for Ecological Recovery, one of Thailand's leading ENGOs. In contrast, Princen and Finger (1994), and Meyer (1995), among others, probe the distinctive organisational traits and qualities of this type of actor.

Grassroots actors

The livelihood struggles of grassroots actors figure prominently in the literature. Ostrom (1990) explains 'traditional' community environmental management practices, while the *Ecologist* (1993) describes the 'tragedy of enclosure' whereby poor farmers, pastoralists, shifting cultivators, fishers and hunter-gatherers were displaced by state-sponsored commercial resource exploitation. Drawing on Scott (1985), Guha (1989), Peluso (1992), Jarosz (1993) and Bryant (1997a) describe the historical and contemporary 'everyday resistance' of these actors. Women are often especially disadvantaged in grassroots livelihood struggles. Shiva (1988), Agarwal (1992) and Jackson (1993) provide contrasting explanations for this situation, while fine studies by Carney (1993), Schroeder (1993), Joekes *et al.* (1995) and Rocheleau *et al.* (1996) elaborate the gendered nature of these struggles. Hong (1987) and Denslow and Padoch (1988), in contrast, show how poor ethnic minority peoples are often victimised as a result of modern 'development'. Yet, as Hecht and Cockburn (1989), Gadgil and Guha (1992), Ghai and Vivian (1992), Schmink and Wood (1992), Broad (1993), and Friedmann and Rangan (1993) illustrate, contemporary conflict usually combines covert resistance with public campaigning, notably using grassroots organisations. Douglass (1992) and Swyngedouw (1995) illustrate that struggles over environmental access are also a fixture in urban political ecology.

BIBLIOGRAPHY

Adams, P. (1991) *Odious Debts: Loose-Lending, Corruption, and the Third World's Environmental Legacy*, Earthscan, London.

Adams, R. (1975) *Energy and Structure: A Theory of Social Power*, University of Texas Press, Austin.

Adams, W.M. (1990) *Green Development: Environment and Sustainability in the Third World*, Routledge, London.

—— (1993) Sustainable development and the greening of development theory, in F.J. Schuurman (ed.), *Beyond the Impasse: New Directions in Development Theory*, Zed Books, London, pp. 207–22.

Adas, M. (1981) From avoidance to confrontation: peasant protest in precolonial and colonial South East Asia, *Comparative Studies in Society and History* 23, 217–47.

—— (1983) Colonization, commercial agriculture, and the destruction of the deltaic rainforests of British Burma in the late nineteenth century, in R.P. Tucker and J.F. Richards (eds), *Global Deforestation and the Nineteenth-Century World Economy*, Duke University Press, Durham, North Carolina, pp. 95–110.

—— (1989) *Machines as the Measure of Men: Science, Technology, and Ideologies of Western Dominance*, Cornell University Press, Ithaca, New York.

Agarwal, A. and Narain, S. (1991) *Global Warming in an Unequal World: A Case of Environmental Colonialism*, Center for Science and Environment, New Delhi.

Agarwal, B. (1988) Under the cooking pot: the political economy of the domestic fuel crisis in rural South Asia, in S. Sontheimer (ed.), *Women and the Environment: A Reader*, Earthscan, London, pp. 93–116.

—— (1992) The gender and environment debate: lessons from India, *Feminist Studies* 18, 119–157.

Allen, E. (1992) Calha Norte: military development in Brazilian Amazonia, *Development and Change* 23, 71–100.

Amanor, K. (1994) *The New Frontier: Farmer Responses to Land Degradation: A West African Study*, Zed Books, London.

Anderson, A.B. (1990) Smokestacks in the rainforest: industrial development and deforestation in the Amazon basin, *World Development* 18, 1191–205.

Anderson, B.R. O'G. (1991) *Imagined Communities: Reflections on the Origin And Spread of Nationalism*, revised edition, Verso, London.

Anderson, R.S. and Huber, W. (1988) *The Hour of the Fox: Tropical Forests, the World Bank, and Indigenous People in Central India*, University of Washington Press, Seattle.

Atkinson, A. (1991) *Principles of Political Ecology*, Belhaven, London.

Axelrod, R. (1984) *The Evolution of Cooperation*, Basic Books, New York.

Bailey, S. (1996) The political ecology of integrated conservation-development projects: case studies from Ecuador, unpublished manuscript.

Banuri, T. and Marglin, F.A. (eds) (1993) *Who Will Save the Forests? Knowledge, Power and Environmental Destruction*, Zed Books, London.

Barrett, C.B. and Arcese, P. (1995) Are integrated conservation-development projects (ICDPs) sustainable? On the conservation of large mammals in Sub-Saharan Africa, *World Development* 23, 1073–84.

Bassett, T.J. (1988) The political ecology of peasant-herder conflicts in the northern Ivory Coast, *Annals of the Association of American Geographers* 78, 453–72.

Bebbington, A., Carrasco, H., Peralbo, L., Ramon, G., Torres, V.H. and Trujilli, J. (1993) Fragile lands, fragile organizations: Indian organizations and the politics of sustainability in Ecuador, *Transactions of the Institute of British Geographers* 18, 179–96.

Beck, U. (1992) *Risk Society: Towards a New Modernity*, Sage, London.

—— (1995) *Ecological Politics in an Age of Risk*, Polity Press, Cambridge.

Beckerman, W. (1974) *In Defence of Economic Growth*, Jonathan Cape, London.

Beinart, W. (1984) Soil erosion, conservationism and ideas about development: a Southern African exploration, *Journal of Southern African Studies* 11, 52–83.

Beinart, W. and Coates, P. (1995) *Environment and History: The Taming of Nature in the USA and South Africa*, Routledge, London.

Benedick, R.E. (1991) *Ozone Diplomacy: New Directions in Safeguarding the Planet*, Harvard University Press, Cambridge, Massachusetts.

Bennett, J.W. (1976) *The Ecological Transition: Cultural Anthropology and Human Adaptation*, Pergamon, Oxford.

Benton, T. (1989) Marxism and natural limits, *New Left Review* 178, 51–86.

Beresford, M. and Fraser, L. (1992) Political economy of the environment in Vietnam, *Journal of Contemporary Asia* 22, 3–19.

Berkes, F. (ed.) (1989) *Common Property Resources: Ecology and Community Based Sustainable Development*, Belhaven, London.

Berry, S. (1989) Social institutions and access to resources, *Africa* 59, 41–55.

Bhabha, H.K. (1994) *The Location of Culture*, Routledge, London.

Binswanger, H.P. (1991) Brazilian policies that encourage deforestation in the Amazon, *World Development* 19, 821–9.

Black, R. (1990) 'Regional political ecology' in theory and practice: a case study from northern Portugal, *Transactions of the Institute of British Geographers* 15, 35–47.

Blaikie, P. (1985) *The Political Economy of Soil Erosion in Developing Countries*, Longman, London.

—— (1988) The explanation of land degradation in Nepal, in J. Ives and D.C. Pitt (eds), *Deforestation: Social Dynamics in Watersheds and Mountain Ecosystems*, Routledge, London, pp. 132–58.

—— (1989) Environment and access to resources in Africa, *Africa* 59, 18–40.

—— (1995a) Changing environments or changing views? A political ecology for developing countries, *Geography* 80, 203–14.

—— (1995b) Understanding environmental issues, in S. Morse and M. Stocking (eds), *People and Environment*, UCL Press, London, pp. 1–30.

Blaikie, P. and Brookfield, H. (1987) *Land Degradation and Society*, Methuen, London.

Blaikie, P. and Unwin, T. (eds) (1988) *Environmental Crises in Developing Countries*, Developing Areas Research Group Monograph no. 5, Institute of British Geographers, London.

Blaikie, P., Cannon, T., Davis, I. and Wisner, B. (1994) *At Risk: Natural Hazards, People's Vulnerability, and Disasters*, Routledge, London.

Blauert, J. and Guidi, M. (1992) Strategies for autochthonous development: two initiatives in rural Oaxaca, Mexico, in D. Ghai and J.M. Vivian (eds), *Grassroots Environmental Action: People's Participation in Sustainable Development*, Routledge, London, pp. 188–220.

Blaut, J.M. (1993) *The Colonizer's Model of the World: Geographical Diffusionism and Eurocentric History*, Guilford Press, London.

Braganza, G.C. (1996) Philippine community-based forest management: options for sustainable development, in M.J.G. Parnwell and R.L. Bryant (eds), *Environmental Change in South-East Asia: People, Politics and Sustainable Development*, Routledge, London, pp. 311–29.

Bramble, B.J. and Porter, G. (1992) Non-governmental organizations and the making of US international environmental policy, in A. Hurrell and B. Kingsbury (eds), *The International Politics of the Environment: Actors, Interests and Institutions*, Clarendon Press, Oxford, pp. 313–53.

Branes, R. (1991) *Institutional and Legal Aspects of the Environment in Latin America, Including the Participation of Non-governmental Organizations in Environmental Management*, Inter-American Development Bank, Washington, DC.

Branford, S. and Glock, O. (1985) *The Last Frontier: Fighting over Land in the Amazon*, Zed Books, London.

Broad, R. (1993) *Plundering Paradise: The Struggle for the Environment in the Philippines*, University of California Press, Berkeley.

—— (1994) The poor and the environment: friends or foes?, *World Development* 22, 811–22.

Bromley, D.W. (1991) *Environment and Economy: Property Rights and Public Policy*, Blackwell, Oxford.

Brookfield, H.C. (1988) The new great age of clearance and beyond, in J.S. Denslow and C. Padoch (eds), *People of the Tropical Rain Forest*, University of California Press, Berkeley, pp. 209–24.

Browder, J.O. (1988) Public policy and deforestation in the Brazilian Amazon, in R. Repetto and M. Gillis (eds), *Public Policies and the Misuse of Forest Resources*, Cambridge University Press, Cambridge, pp. 247–98.

Brown, R. and Daniel, P. (1991) Environmental issues in mining and petroleum contracts, *IDS Bulletin* 22 (4), 45–9.

Bryant, R.L. (1991) Putting politics first: the political ecology of sustainable development, *Global Ecology and Biogeography Letters* 1, 164–6.

—— (1992) Political ecology: an emerging research agenda in Third-World studies, *Political Geography* 11, 12–36.

—— (1994a) Shifting the cultivator: the politics of teak regeneration in colonial Burma, *Modern Asian Studies* 28, 225–50.

—— (1994b) From laissez-faire to scientific forestry: forest management in early colonial Burma 1826–85, *Forest and Conservation History* 38, 160–70.

—— (1994c) Fighting over the forests: political reform, peasant resistance and the transformation of forest management in late colonial Burma, *Journal of Commonwealth and Comparative Politics* 32, 244–60.

—— (1996a) Romancing colonial forestry: the discourse of 'forestry as progress' in British Burma, *Geographical Journal* 162, 169–78.

—— (1996b) Asserting sovereignty through natural resource use: Karen forest management on the Thai-Burmese border, in R. Howitt, J. Connell and P. Hirsch (eds), *Resources, Nations and Indigenous Peoples: Case Studies from Australasia, Melanesia and South-East Asia*, Oxford University Press, Melbourne, pp. 32–41.

—— (1997a) *The Political Ecology of Forestry in Burma, 1824–1994*, C. Hurst, London and University of Hawaii Press, Honolulu.

—— (1997b) Beyond the impasse: the power of political ecology in Third World environmental research, *Area* 29, 1–15.

Bryant, R.L., Rigg, J. and Stott, P. (eds) (1993) The political ecology of Southeast Asia's forests: trans-disciplinary discourses, Special Issue, *Global Ecology and Biogeography Letters* 3 (4–6), 101–296.

Bryant, R.L. and Parnwell, M.J.G. (1996) Introduction: politics, sustainable development and environmental change in South-East Asia, in M.J.G. Parnwell and R.L. Bryant (eds), *Environmental Change in South-East Asia: People, Politics and Sustainable Development*, Routledge, London, pp. 1–20.

Buchanan, K. (1973) The white north and the population explosion, *Antipode* 3, 7–15.

Bull, D. (1982) *A Growing Problem: Pesticides and the Third-World Poor*, Oxfam, Oxford.

Bullard, R.D. (ed.) (1993) *Confronting Environmental Racism: Voices from the Grassroots*, South End Press, Boston.

Bunker, S.G. (1985) *Underdeveloping the Amazon: Extraction, Unequal Exchange, and the Failure of the Modern State*, University of Illinois Press, Urbana.

Buttel, F.H. (1996) Environmental and resource sociology: theoretical issues and opportunities for synthesis, *Rural Sociology* 61, 56–76.

Caldwell, L.K. (1990) *International Environmental Policy: Emergence and Dimensions*, second edition, Duke University Press, Durham, North Carolina.

Caldwell, M. (1977) *The Wealth of Some Nations*, Zed Press, London.

Carney, J. (1993) Converting the wetlands, engendering the environment: the intersection of gender with agrarian change in The Gambia, *Economic Geography* 69, 329–48.

Carroll, T. (1992) *Intermediary NGOs: The Supporting Link in Grassroots Development*, Kumarian Press, West Hartford, Connecticut.

Carson, R. (1962) *Silent Spring*, Houghton Mifflin, Boston.

Cartwright, J. (1989) Conserving nature, decreasing debt, *Third World Quarterly* 11, 114–27.

Castree, N. (1995) The nature of produced nature: materiality and knowledge construction in Marxism, *Antipode* 27, 12–48.

Centre for Science and Environment (CSE) (1985) *The State of India's Environment*, CSE, New Delhi.

—— (1992) The Centre for Science and Environment statement on global environmental democracy, *Alternatives* 17, 261–79.

Chambers, R. (1983) *Rural Development: Putting the Last First*, Longman, London.

—— (1987) *Sustainable Livelihoods, Environment and Development: Putting Poor Rural People First*, Institute of Development Studies discussion paper no. 240, University of Sussex, Brighton.

Chapman, M.D. (1989) The political ecology of fisheries depletion in Amazonia, *Environmental Conservation* 16, 331–7.

Chatterjee, P. and Finger, M. (1994) *The Earth Brokers: Power, Politics and World Development*, Routledge, London.

Clark, J. (1991) *Democratizing Development: The Role of Voluntary Organisations*, Earthscan, London.

—— (1995) The state, popular participation, and the voluntary sector, *World Development* 23, 593–601.

Clarke, W.C. and Munn, R.E. (eds) (1986) *Sustainable Development of the Biosphere*, Cambridge University Press, Cambridge.

Cleary, D. (1990) *Anatomy of the Amazon Gold Rush*, Macmillan, London.

Cliffe, L. and Moorsom, R. (1979) Rural class formation and ecological collapse in Botswana, *Review of African Political Economy* 15/16, 35–52.

Cline, W. (1982) Can the East Asian model of development be generalized?, *World Development* 10, 81–90.

Cochrane, J. (1996) The sustainability of ecotourism in Indonesia: fact and fiction, in M.J.G. Parnwell and R.L. Bryant (eds), *Environmental Change in South-East Asia: People, Politics and Sustainable Development*, Routledge, London, pp. 237–59.

Cockburn, A. and Ridgeway, J. (eds) (1979) *Political Ecology*, Times Books, New York.

Colburn, F.D. (ed.) (1989) *Everyday Forms of Peasant Resistance*, M.E. Sharpe, London.

Colchester, M. (1993) Pirates, squatters and poachers: the political ecology of dispossession of the native peoples of Sarawak, *Global Ecology and Biogeography Letters* 3 (4–6), 158–79.

—— (1994a) Sustaining the forests: the community-based approach in South and South-East Asia, *Development and Change* 25, 69–100.

—— (1994b) The new sultans: Asian loggers move in on Guyana's forests, *The Ecologist* 24, 45–52.

Colchester, M. and Lohmann, L. (1990) *The Tropical Forestry Action Plan: What Progress?*, The Ecologist, Sturminster Newton, Dorset.

Collett, D. (1987) Pastoralists and wildlife: image and reality in Kenya Maasailand, in D. Anderson and R. Grove (eds), *Conservation in Africa: Peoples, Policies and Practice*, Cambridge University Press, Cambridge, pp. 129–49.

Collins, R.O. (1990) *The Waters of the Nile: Hydropolitics and the Jonglei Canal, 1900–1988*, Clarendon, Oxford.

Commoner, B. (1972) *The Closing Circle*, Bantam Books, New York.

Conroy, C. and Litvinoff, M. (eds) (1988) *The Greening of Aid: Sustainable Livelihoods in Practice*, Earthscan, London.

Constantino-David, K. (1995) Community organizing in the Philippines: the experience of development NGOs, in G. Craig and M. Mayo (eds), *Community Empowerment: A Reader in Participation and Development*, Zed Books, London, pp. 154–67.

Coote, B. (1995) *NAFTA: Poverty and Free Trade in Mexico*, Oxfam, Oxford.

Corbridge, S. (1986) *Capitalist World Development: A Critique of Radical Development Geography*, Macmillan, London.

Cosgrove, D. and Daniels, S.J. (eds) (1988) *The Iconography of Landscape: Essays on the Symbolic Representation, Design and Use of Past Environments*, Cambridge University Press, Cambridge.

Cox, R. (1987) *Production, Power and World Order*, Columbia University Press, New York.

Croll, E. and Parkin, D. (1992) Cultural understandings of the environment, in E. Croll and D. Parkin (eds), *Bush Base, Forest Farm: Culture, Environment and Development*, Routledge, London, pp. 11–36.

Cronon, W. (ed.) (1995) *Uncommon Ground: Toward Reinventing Nature*, W.W. Norton, New York.

Cummings, B.J. (1990) *Dam the Rivers, Damn the People: Development and Resistance in Amazonian Brazil*, Earthscan, London.

Dalby, S. (1992) Ecopolitical discourse: 'environmental security' and political geography, *Progress in Human Geography* 16, 503–22.

Dankelman, I. and Davidson, J. (1988) *Women and Environment in the Third World: Alliance for the Future*, Earthscan, London.

Darden, J. (1975) Population control or a redistribution of wealth?, *Antipode* 7, 50–2.

Dauvergne, P. (1993/4) The politics of deforestation in Indonesia, *Pacific Affairs* 66, 497–518.

Davidson, J., Chakraborty, M. and Myers, D. (1992) *No Time to Waste: Poverty and the Global Environment*, Oxfam, Oxford.

Davies, S. (1992) Green conditionality and food security: winners and losers from the greening of aid, *Journal of International Development* 4, 151–65.

Denslow, J.S. and Padoch, C. (eds) (1988) *People of the Tropical Rain Forest*, University of California Press, Berkeley.

Deutsch, K.W. (1977) Some problems and prospects of ecopolitical research, in K.W. Deutsch (ed.), *Ecosocial Systems and Ecopolitics: A Reader on Human and Social Implications of Environmental Management in Developing Countries*, UNESCO, Paris, pp. 359–68.

Diegues, A.C.S. (1992) Sustainable development and people's participation in wetland ecosystem conservation in Brazil: two comparative studies, in D. Ghai and J.M. Vivian (eds), *Grassroots Environmental Action: People's Participation in Sustainable Development*, Routledge, London, pp. 141–59.

Dinham, B. (1991) FAO and pesticides: promotion or proscription?, *The Ecologist* 21, 61–5.

Dinham, B. and Hines, C. (1983) *Agribusiness in Africa*, Earth Resources Research, London.

Dobson, A. (1995) *Green Political Thought*, second edition, Routledge, London.

Doedens, A., Persoon, G. and Wedda, C. (1995) The relevance of ethnicity in the depletion and management of forest resources in Northeast Luzon, Philippines, *Sojourn* 10, 259–79.

Dore, E. (1996) Capitalism and ecological crisis: legacy of the 1980s, in H. Collinson (ed.), *Green Guerrillas: Environmental Conflicts and Initiatives in Latin America and the Caribbean*, Latin America Bureau, London, pp. 8–19.

Douglass, M. (1992) The political economy of urban poverty and environmental management in Asia: access, empowerment and community based alternatives, *Environment and Urbanisation* 4, 9–32.

Dryzek, J.S. (1987) *Rational Ecology: Environment and Political Economy*, Basil Blackwell, Oxford.

Durning, A.T. (1990) Ending poverty, in L.R Brown (ed.), *State of the World 1990*, Unwin Hyman, London, pp. 135–53.

—— (1993) Supporting indigenous peoples, in L.R. Brown (ed.), *State of the World 1993*, Earthscan, London, pp. 80–100.

Dwivedi, O.P. and Vajpeyi, D.K. (eds) (1995) *Environmental Policies in the Third World*, Greenwood Press, Wesport, Connecticut.

Eccleston, B. (1996) Does North–South collaboration enhance NGO influence on deforestation policies in Malaysia and Indonesia?, *Journal of Commonwealth and Comparative Politics* XXXIV, 66–89.

Eccleston, B. and Potter, D. (1996) Environmental NGOs and different political contexts in South-East Asia: Malaysia, Indonesia and Vietnam, in M.J.G. Parnwell and R.L. Bryant (eds), *Environmental Change in South-East Asia: People, Politics and Sustainable Development*, Routledge, London, pp. 49–66.

Ecologist, The (1993) *Whose Common Future? Reclaiming the Commons*, Earthscan, London.

Eden, S.E. (1994) Using sustainable development: the business case, *Global Environmental Change* 4, 160–7.

Ehrlich, P.R. (1968) *The Population Bomb*, Ballantine, London.

Ehrlich, P. and Ehrlich, A. (1990) *The Population Explosion*, Hutchinson, London.

Ekins, P. (1992) *A New World Order: Grassroots Movements for Global Change*, Routledge, London.

Eldridge, P.J. (1995) *Non-government Organizations and Democratic Participation in Indonesia*, Oxford University Press, Kuala Lumpur.

Ellen, R.F. (1982) *Environment, Subsistence and System: The Ecology of Small-scale Social Formations*, Cambridge University Press, Cambridge.

Elliott, J.A. (1994) *An Introduction to Sustainable Development: The Developing World*, Routledge, London.

Enzensberger, H.M. (1974) A critique of political ecology, *New Left Review* 84, 3–31.

Escobar, A. (1988) Power and visibility: development and the invention and management of the Third World, *Cultural Anthropology* 3, 428–43.

—— (1995) *Encountering Development: The Making and Unmaking of the Third World*, Princeton University Press, Princeton.

—— (1996) Constructing nature: elements for a poststructural political ecology, in R. Peet and M. Watts (eds), *Liberation Ecologies: Environment, Development, Social Movements,* Routledge, London, pp. 46–68.

Esteva, G. and Prakash, M.S. (1992) Grassroots resistance to sustainable development, *The Ecologist* 22, 45–51.

Eyre, S.R. and Jones, G.R.J. (1967) *Geography as Human Ecology*, Edward Arnold, London.

Fairhead, J. and Leach, M. (1995) False forest history, complicit social analysis: rethinking some West African environmental narratives, *World Development* 23, 1023–35.

Fairlie, S. (ed.) (1995) Overfishing: its causes and consequences, Special Issue, *The Ecologist* 25 (2/3), 42–125.

Farrington, J. and Bebbington, A. (1993) *Reluctant Partners? Non-governmental Organisations, the State and Sustainable Agricultural Development*, Routledge, London.

Ferguson, J. (1990) *The Anti-politics Machine: Development, Depoliticization and Bureaucratic Power in the Third World*, Cambridge University Press, Cambridge.

Fisher, J. (1993) *The Road from Rio: Sustainable Development and the Non-governmental Movement in the Third World*, Praeger, Wesport, Connecticut.

Food and Agriculture Organisation (FAO), World Resources Institute, World Bank and UN Development Program (1987) *The Tropical Forestry Action Plan*, FAO, Rome.

Forsyth, T. (1996) Science, myth, and knowledge: testing Himalayan environmental degradation in Thailand, *Geoforum* 27, 375–92.

Fortmann, L. (1995) Talking claims: discursive strategies in contesting property, *World Development* 23, 1053–63.

Foucault, M. (1977) *Discipline and Punish: The Birth of the Prison*, Pantheon Books, New York.

Frank, A.G. (1967) *Capitalism and Underdevelopment in Latin America*, Monthly Review Press, New York.

Franke, R.W. and Chasin, B.H. (1980) *Seeds of Famine: Ecological Destruction and the Development Dilemma in the West African Sahel*, Allanheld Osmun, Montclair, New Jersey.

Friedmann, J. (1992) *Empowerment: The Politics of Alternative Development*, Blackwell, Oxford.

Friedmann, J. and Rangan, H. (eds) (1993) *In Defense of Livelihood: Comparative Studies in Environmental Action*, Kumarian Press, West Hartford, Connecticut.

Friends of the Earth (1992) *The Rainforest Harvest: Sustainable Strategies for Saving the Tropical Forests?*, Friends of the Earth, London.

—— (1993) *Mahogany is Murder*, Friends of the Earth, London.

—— (1994) *Crude Operator: The Environmental, Social and Cultural Effects of Texaco Oil Operations in the Tropical Forests of Ecuador*, Friends of the Earth, London.

Furnivall, J.S. (1956) *Colonial Policy and Practice: A Comparative Study of Burma and Netherlands India*, New York University Press, New York.

Gadgil, M. and Guha, R. (1992) *This Fissured Land: An Ecological History of India*, Oxford University Press, Delhi.

—— (1994) Ecological conflicts and the environmental movement in India, *Development and Change* 25, 101–36.

Gan, L. (1993) The making of the Global Environmental Facility: an actor's perspective, *Global Environmental Change* 3, 256–75.

Garner, R. (1996) *Environmental Politics*, Harvester Wheatsheaf, London.

George, S. (1988) *A Fate Worse than Debt*, Penguin, Harmondsworth.

—— (1992) *The Debt Boomerang*, Pluto, London.

George, S. and Sabelli, F. (1994) *Faith and Credit: The World Bank's Secular Empire*, Penguin, Harmondsworth.

Ghai, D. and Vivian, J.M. (eds) (1992) *Grassroots Environmental Action: People's Participation in Sustainable Development*, Routledge, London.

Ghimire, K.B. (1994) Parks and people: livelihood issues in national parks management in Thailand and Madagascar, *Development and Change* 25, 195–229.

Giddens, A. (1979) *Central Problems in Social Theory: Action, Structure Contradiction in Social Analysis*, Macmillan, London.

Gill, S. and Law, D. (1988) *The Global Political Economy*, Harvester Wheatsheaf, London.

Gillis, M. (1988) Indonesia: public policies, resource management, and the tropical forest, in R. Repetto and M. Gillis (eds), *Public Policies and the Misuse of Forest Resources*, Cambridge University Press, Cambridge, pp. 43–114.

Gilpin, R. (1987) *The Political Economy of International Relations*, Princeton University Press, Princeton, New Jersey.

Gladwin, T.N. (1987) Environment, development and multinational enterprise, in C.S. Pearson (ed.), *Multinational Corporations, Environment, and the Third World: Business Matters*, Duke University Press, Durham, North Carolina, pp. 3–31.

Goldsmith, E. and Hildyard, N. (1984) *The Social and Environmental Effects of Large Dams*, vol. 1, Wadebridge Ecological Centre, Wadebridge, Cornwall.

—— (eds) (1986) *The Social and Environmental Effects of Large Dams*, vol. 2, Wadebridge Ecological Centre, Wadebridge, Cornwall.

—— (1991) World agriculture – Toward 2000: FAO's plan to feed the world, *The Ecologist* 21, 81–91.

Goudie, A. (1993) *The Human Impact on the Natural Environment*, fourth edition, Blackwell, Oxford.

Greider, T. and Garkovich, L. (1994) Landscapes: the social construction of nature and the environment, *Rural Sociology* 59, 1–24.

Grossman, L.S. (1984) *Peasants, Subsistence Ecology and Development in the Highlands of Papua New Guinea*, Princeton University Press, Princeton.

—— (1993) The political ecology of banana exports and local food production in St Vincent, Eastern Caribbean, *Annals of the Association of American Geographers* 83, 347–67.

Grove, R.H. (1990) Colonial conservation, ecological hegemony and popular resistance: towards a global synthesis, in J.M. MacKenzie (ed.), *Imperialism and the Natural World*, Manchester University Press, Manchester, pp. 15–50.

Grundemann, R. (1991) *Marxism and Ecology*, Oxford University Press, Oxford.

Guha, R. (1989) *The Unquiet Woods: Ecological Change and Peasant Resistance in the Himalaya*, Oxford University Press, Delhi.

Guimaraes, R.P. (1991) *The Eco-politics of Development in the Third World: Politics and Environment in Brazil*, Lynne Rienner, Boulder.

Gupta, S. *et al.* (1995) Public expenditure policy and the environment: a review and synthesis, *World Development* 23, 515–28.

Guyer, J.L. and Peters, P.E. (eds) (1987) Conceptualizing the household: issues of theory and policy in Africa, Special issue, *Development and Change* 18 (2).

Hall, A.L. (1989) *Developing Amazonia: Deforestation and Social Conflict in Brazil's Carajas Programme*, Manchester University Press, Manchester.

Hall, J.A. (ed.) (1986) *States in History*, Basil Blackwell, Oxford.

Handley, P. (1994) Parks under siege: Thai plan to develop national parks for tourism faces opposition, *Far Eastern Economic Review* (20 January 1994), 36–7.

Hardesty, D.L. (1977) *Ecological Anthropology*, Wiley, New York.

Hardin, G. (1968) The tragedy of the commons, *Science* 162, 1243–8.

Hardin, G. and Baden, J. (eds) (1977) *Managing the Commons*, W.H. Freeman, San Francisco.

Hardoy, J.E., Mitlin, D. and Satterthwaite, D. (1992) *Environmental Problems in Third World Cities*, Earthscan, London.

Harris, N. (1986) *The End of the Third World: Newly Industrializing Countries and the Decline of an Ideology*, Penguin, Harmondsworth.

Harrison, P. (1993) *The Third Revolution: Population, Environment and a Sustainable World*, Penguin, Harmondsworth.

Harvey, D. (1974) Population, resources and the ideology of science, *Economic Geography* 50, 256–77.

—— (1993) The nature of environment: the dialectics of social and environmental change, in R. Miliband and L. Panitch (eds), *Real Problems, False Solutions: Socialist Register 1993*, Merlin Press, London, pp. 1–51.

Hayter, T. (1989) *Exploited Earth: Britain's Aid and the Environment: A Friends of the Earth Inquiry*, Earthscan, London.

Hayward, T. (1995) *Ecological Thought: An Introduction*, Polity Press, Cambridge.

Headrick, D.R. (1981) *The Tools of Empire: Technology and European Imperialism in the Nineteenth Century*, Oxford University Press, Oxford.

Hecht, S.B. (1985) Environment, development and politics: capital accumulation and the livestock sector in eastern Amazonia, *World Development* 13, 663–84.

Hecht, S.B. and Cockburn, A. (1989) *The Fate of the Forests: Developers, Destroyers and Defenders of the Amazon*, Verso, London.

—— (1992) Realpolitik, reality and rhetoric in Rio, *Environment and Planning D: Society and Space* 10, 367–75.

Hecht, S.B., Anderson, A.B. and May, P. (1988) The subsidy from nature: shifting cultivation, successional palm forests, and rural development, *Human Organization* 47, 25–35.

Hedlund, H. (1979) Contradictions in the peripheralization of a pastoral society: the Maasai, *Review of African Political Economy* 15/16, 53–62.

Heilbroner, R. (1974) *An Inquiry into the Human Prospect*, Norton, New York.

Hershkovitz, L. (1993) Political ecology and environmental management in the Loess Plateau, China, *Human Ecology* 21, 327–53.

Hettne, B. (1995) *Development Theory and the Three Worlds*, second edition, Longman, London.

Hildyard, N. (1991) Sustaining the hunger machine: a critique of FAO's sustainable agriculture and rural development strategy, *The Ecologist* 21, 239–43.

210

Hill, S. (1988) *The Tragedy of Technology*, Pluto Press, London.

Hirsch, P. (1990) *Development Dilemmas in Rural Thailand*, Oxford University Press, Singapore.

—— (1995) A state of uncertainty: political economy of community resource management at Tab Salao, Thailand, *Sojourn* 10, 172–97.

Hirsch, P. and Lohmann, L. (1989) Contemporary politics of environment in Thailand, *Asian Survey* 29, 439–51.

Hjort, A. (1982) A critique of 'ecological' models of pastoral land-use, *Nomadic Peoples* 10, 11–27.

Hobbes, T. (1968) *Leviathan*, Penguin, Harmondsworth.

Hoben, A. (1995) Paradigms and politics: the cultural construction of environmental policy in Ethiopia, *World Development* 23, 1007–21.

Hobsbawm, E.J. (1996) The future of the state, *Development and Change* 27, 267–78.

Hoffert, R.W. (1986) The scarcity of politics: Ophuls and Western political thought, *Environmental Ethics* 8, 5–32.

Hong, E. (1987) *Natives of Sarawak: Survival in Borneo's Vanishing Forests*, Institut Masyarakat, Penang.

Horowitz, M.M. and Little, P.D. (1987) African pastoralism and poverty: some implications for drought and famine, in M.H. Glantz (ed.), *Drought and Hunger in Africa: Denying Famine a Future*, Cambridge University Press, Cambridge, pp. 59–82.

Hurrell, A. (1992) Brazil and the international politics of Amazonian deforestation, in A. Hurrell and B. Kingsbury (eds), *The International Politics of the Environment: Actors, Interests, and Institutions*, Clarendon Press, Oxford, pp. 398–429.

—— (1994) A crisis of ecological viability? Global environmental change and the nation state, *Political Studies* 42, 146–65.

Hurrell, A. and Kingsbury, B. (eds) (1992) *The International Politics of the Environment: Actors, Interests, and Institutions,* Clarendon Press, Oxford.

Hurst, P. (1990) *Rainforest Politics: Ecological Destruction in South-East Asia*, Zed Books, London.

Ingold, T. (1992) Culture and the perception of the environment, in E. Croll and D. Parkin (eds), *Bush Base, Forest Farm: Culture, Environment and Development*, Routledge, London, pp. 39–56.

IUCN, UNEP and WWF (1991) *Caring for the Earth: A Strategy for Sustainable Living*, Earthscan, London.

Ives, J. and Messerli, B. (1989) *The Himalyan Dilemma: Reconciling Development and Conservation*, Routledge, London.

Jackson, C. (1993) Environmentalisms and gender interests in the Third World, *Development and Change* 24, 649–77.

Jacobson, J.L. (1988) *Environmental Refugees: A Yardstick of Habitability*, Worldwatch Paper no. 86, Worldwatch Institute, Washington, DC.

Jain, S. (1991) Standing up for trees: women's role in the Chipko movement, in S. Sontheimer (ed.), *Women and the Environment: A Reader*, Earthscan, London, pp. 163–78.

Jarosz, L. (1993) Defining and explaining tropical deforestation: shifting cultivation and population growth in colonial Madagascar (1896–1940), *Economic Geography* 69, 366–79.

Jewitt, S. (1995) Europe's 'Others'? Forestry policy and practice in colonial and postcolonial India, *Environment and Planning D: Society and Space* 13, 67–90.

Joekes, S., Leach, M. and Green, C. (eds) (1995) Gender relations and environmental change, Special Issue, *IDS Bulletin* 26, 1–95.

Johnston, R.J. (1989) *Environmental Problems: Nature, Economy and State*, Belhaven, London.

—— (1992) Laws, states and super-states: international law and the environment, *Applied Geography* 12, 211–28.

Juniper, T. (1992) *Whose Hand on the Chainsaw? UK Government Policy and the Tropical Rainforests*, Friends of the Earth, London.

Kendie, S.B. (1995) The environmental dimensions of structural adjustment programmes: missing links to sustaining development, *Singapore Journal of Tropical Geography* 16, 42–57.

Keohane, R.O. (ed.) (1986) *Neorealism and its Critics*, Columbia University Press, New York.

Kimerling, J. (1996) Oil, lawlessness and indigenous struggles in Ecuador's Oriente, in H. Collinson (ed.), *Green Guerrillas: Environmental Conflicts and Initiatives in Latin America and the Caribbean*, Latin America Bureau, London, pp. 61–73.

King, V.T. (1993) Politik pembangunan: the political economy of rainforest exploitation and development in Sarawak, East Malysia, *Global Ecology and Biogeography Letters* 3, 235–44.

Kloppenburg, J.R. (1988) *First the Seed: The Political Economy of Plant Biotechnology, 1492–2000*, Cambridge University Press, Cambridge.

Kloppenburg, J. and Burrows, B. (1996) Biotechnology to the rescue? Twelve reasons why biotechnology is incompatible with sustainable agriculture, *The Ecologist* 26, 61–7.

Kneen, B. (1995) The invisible giant: Cargill and its transnational strategies, *The Ecologist* 25, 195–9.

Korten, D.C. (1990) *Getting to the 21st Century: Voluntary Action and the Global Agenda*, Kumarian Press, West Hartford, Connecticut.

—— (1995) *When Corporations Rule the World*, Earthscan, London.

Kothari, A., Suri, S. and Singh, N. (1995) People and protected areas: rethinking conservation in India, *The Ecologist* 25, 188–94.

Krasner, S.D. (1985) *Structural Conflict: The Third World against Global Liberalism*, University of California Press, Berkeley.

Kumar, N. (1993) Biotechnologies and sustainable development, in M.C. Howard (ed.), *Asia's Environmental Crisis*, Westview, Boulder, Colorado, pp. 169–79.

Lane, C. (1992) The Barabaig pastoralists of Tanzania: sustainable land use in jeopardy, in D. Ghai and J.M. Vivian (eds), *Grassroots Environmental Action: People's Participation in Sustainable Development*, Routledge, London, pp. 81–105.

Leach, M. (1991) Engendered environments: understanding natural resource management in the West African forest zone, *IDS Bulletin* 22 (4), 17–24.

—— (1993) Women's use of forest resources in Sierra Leone, in A.Rodda (ed.), *Women and the Environment*, Zed Books, London, pp. 126–9.

Legazpi, E. (1994) Environmental coalitions, in C.P. Cala and J.Z. Grageda (eds), *Studies on Coalition Experiences in the Philippines*, Bookmark, Manila, pp. 121–54.

Leiberman, V.B. (1984) *Burmese Administrative Cycles: Anarchy and Conquest, c. 1580–1760*, Princeton University Press, Princeton.

Leitmann, J., Bartone, C. and Bernstein, J. (1992) Environmental management and urban development: issues and options for Third World cities, *Environment and Urbanization* 4, 131–40.

Leonard, H.J. (1988) *Pollution and the Struggle for the World Product: Multinational Corporations, Environment, and International Comparative Advantage*, Cambridge University Press, Cambridge.

Le Prestre, P.G. (1989) *The World Bank and the Environmental Challenge*, Associated University Presses, London.

Leungaramsri, P. and Rajesh, N. (1992) *The Future of People and Forests in Thailand after the Logging Ban*, Project for Ecological Recovery, Bangkok.

Lewis, M.W. (1992) *Green Delusions: An Environmentalist Critique of Radical Environmentalism*, Duke University Press, Durham, North Carolina.

Lipschutz, R.D. and Conca, K. (eds) (1993) *The State and Social Power in Global Environmental Politics*, Columbia University Press, New York.

Litfin, K.T. (1994) *Ozone Discourses: Science and Politics in Global Environmental Change*, Columbia University Press, New York.

Little, P.D. and Horowitz, M.M. (eds) (1987) *Lands at Risk in the Third World: Local-Level Perspectives*, Westview Press, Boulder, Colorado.

Little, P.D. and Watts, M. (eds) (1994) *Living Under Contract: Contract Farming and Agrarian Transformation in sub-Saharan Africa*, University of Wisconsin Press, Wisconsin.

Livernash, R. (1992) Policies and institutions – non-governmental organizations: a growing force in the developing world, in World Resources Institute (ed.), *World Resources 1992–93*, Oxford University Press, Oxford, pp. 215–34.

Lohmann, L. (1993) Land, power and forest colonization in Thailand, *Global Ecology and Biogeography Letters* 3, 180–91.

—— (1996) Freedom to plant: Indonesia and Thailand in a globalising pulp and paper industry, in M.J.G. Parnwell and R.L. Bryant (eds), *Environmental Change in South-East Asia: People, Politics and Sustainable Development*, Routledge, London, pp. 23–48.

Long, N. and Long, A. (eds) (1992) *Battlefields of Knowledge: The Interlocking of Theory and Practice in Social Research and Development*, Routledge, London.

Lowe, P. and Worboys, M. (1978) Ecology and the end of ideology, *Antipode* 10, 12–21.

Luke, S. (1977) *Power: A Radical View*, Macmillan, London.

MacAndrews, C. (1994) The Indonesian Environmental Impact Management Agency (BAPEDAL): its role, development and future, *Bulletin of Indonesian Economic Studies* 30, 85–103.

MacKenzie, J.M. (1988) *The Empire of Nature: Hunting, Conservation and British Imperialism*, Manchester University Press, Manchester.

McCay, B.J. and Acheson, J.M. (eds) (1987) *The Question of the Commons: The Culture and Ecology of Communal Resources*, University of Arizona Press, Tucson.

McCormick, J. (1995) *The Global Environmental Movement: Reclaiming Paradise*, second edition, Wiley, Chichester.

McD.Beckles, H. (1996) Where will all the garbage go? Tourism, politics, and the environment in Barbados, in H. Collinson (ed.), *Green Guerrillas: Environmental Conflicts and Initiatives in Latin America and the Caribbean*, Latin America Bureau, London, pp. 187–93.

McDonald, M.D. (1993) Dams, displacement and development: a resistance movement in southern Brazil, in J. Friedmann and H. Rangan (eds), *In Defense of Livelihood: Comparative Studies in Environmental Action*, Kumarian Press, West Hartford, Connecticut, pp. 79–105.

McDowell, M.A. (1989) Development and the environment in ASEAN, *Pacific Affairs* 62, 307–29.

McKibbens, B. (1989) *The End of Nature*, Viking, New York.

Madeley, J. (1994) Green Revolution blues, *Perspectives* 13, 16–17.

Mahoney, R. (1992) Debt-for-nature swaps – who really benefits?, *The Ecologist* 22, 97–103.

Mann, M. (1984) The autonomous power of the state: its origins, mechanisms and results, *Archives européennes de sociologie* 25, 185–213.

BIBLIOGRAPHY

—— (1986) *The Sources of Social Power I: A History of Power from the Beginning to AD 1760*, Cambridge University Press, Cambridge.

Marchak, M.P. (1995) *Logging the Globe*, McGill-Queen's University Press, Montreal and Kingston.

Marshall, G. (1991) FAO and tropical forestry, *The Ecologist* 21, 66–72.

Martinez-Alier, J. (1990) *Ecological Economics: Energy, Environment and Society*, Blackwell, Oxford.

Meadows, D., Randers, J. and Behrens, W.W. (1972) *The Limits to Growth*, Universe Books, New York.

Meyer, C.A. (1995) Opportunism and NGOs: entrepreneurship and green North–South transfers, *World Development* 23, 1277–89.

Meyer-Abich, K.M. (1993) Winners and losers in climate change, in W. Sachs (ed.), *Global Ecology: A New Arena of Political Conflict*, Zed Books, London, pp. 68–87.

Middleton, N., O'Keefe, P. and Mayo, S. (1993) *The Tears of the Crocodile: From Rio to Reality in the Developing World*, Pluto Press, London.

Mies, M. and Shiva, V. (1993) *Ecofeminism*, Zed Books, London.

Migdal, J. (1988) *Strong Societies and Weak States: State–Society Relations and State Capabilities in the Third World*, Princeton University Press, Princeton.

Mikesell, R.F. and Williams, L. (1992) *International Banks and the Environment: From Growth to Sustainability, and Unfinished Agenda*, Sierra Club, San Francisco.

Miller, M.A.L. (1995) *The Third World in Global Environmental Politics*, Open University Press, Buckingham.

Miller, S.W. (1994) Fuelwood in colonial Brazil: the economic and social consequences of fuel depletion for the Bahian Reconcavo, 1549–1820, *Forest and Conservation History* 38, 181–92.

Mische, P.M. (1989) Ecological security and the need to reconceptualize sovereignty, *Alternatives* 14, 389–427.

Moguel, J. and Velazquez, E. (1992) Urban social organization and ecological struggle in Drango, Mexico, in D. Ghai and J.M. Vivian (eds), *Grassroots Environmental Action: People's Participation in Sustainable Development*, Routledge, London, pp. 161–87.

Momtaz, D. (1996) The United Nations and the protection of the environment: from Stockholm to Rio de Janeiro, *Political Geography* 15, 261–71.

Moody, R. (1996) Mining the world: the global reach of Rio Tinto Zinc, *The Ecologist* 26, 46–52.

Moore, D.S. (1993) Contesting terrain in Zimbabwe's Eastern Highlands: political ecology, ethnography, and peasant resource struggles, *Economic Geography* 69, 380–401.

—— (1996) Marxism, culture and political ecology: environmental struggles in Zimbabwe's Eastern Highlands, in R. Peet and M. Watts (eds), *Liberation ecologies: environment, development, social movements*, Routledge, London, pp. 125–47.

Morehouse, W. (1994) Unfinished business: Bhopal ten years after, *The Ecologist* 24, 164–8.

Mortimore, M. (1989) *Adapting to Drought: Farmers, Famines and Desertification in West Africa*, Cambridge University Press, Cambridge.

Muldavin, J.S.S. (1996) The political ecology of agrarian reform in China, in R. Peet and M. Watts (eds), *Liberation Ecologies: Environment, Development, Social Movements*, Routledge, London, pp. 227–59.

Murdoch, J. and Clark, J. (1994) Sustainable knowledge, *Geoforum* 25, 115–32.

Murphy, C.N. and Tooze, R. (eds) (1991) *The New International Political Economy*, Lynne Rienner, Boulder, Colorado.

Neumann, R.P. (1992) Political ecology of wildlife conservation in the Mt Meru area of northeast Tanzania, *Land Degradation and Rehabilitation* 3, 85–98.

—— (1995) Local challenges to global agendas: conservation, economic liberalization and the pastoralists' rights movement in Tanzania, *Antipode* 27, 363–82.

Neumann, R.P. and Schroeder, R.A. (eds) (1995) Manifest ecological destinies, Special Issue, *Antipode* 27, 321–428.

Norgaard, R.B. (1994) *Development Betrayed: The End of Progress and a Coevolutionary Revisioning of the Future*, Routledge, London.

O'Brien, J. (1985) Sowing the seeds of famine: the political economy of food deficits in Sudan, *Review of African Political Economy* 33, 23–32.

O'Connor, J. (1988) Capitalism, nature, socialism: a theoretical introduction, *Capitalism, Nature, Socialism* 1, 11–38.

—— (1989) Uneven and combined development and ecological crisis: a theoretical introduction, *Race and Class* 30, 1–12.

O'Connor, M. (ed.) (1994a) *Is Capitalism Sustainable? Political Economy and the Politics of Ecology*, Guilford Press, London.

—— (1994b) Introduction: liberate, accumulate – and bust?, in M. O'Connor (ed.), *Is Capitalism Sustainable? Political Economy and the Politics of Ecology*, Guilford Press, London, pp. 1–22.

O'Keefe, P. (1975) *African Drought: A Review*, Disaster Research Unit, occasional paper no. 8, University of Bradford.

O'Keefe, P., Westgate, K. and Wisner, B. (1977) Taking the naturalness out of natural disasters, *Nature* 260, 566–7.

Ophuls, W. (1977) *Ecology and the Politics of Scarcity: Prologue to a Political Theory of the Steady State*, W.H. Freeman and Company, San Francisco.

O'Riordan, T. (ed.) (1995) *Environmental Science for Environmental Management*, Longman, London.

Orlove, B.S. (1980) Ecological anthropology, *Annual Review of Anthropology* 9, 235–73.

Orr, D.W. and Hill, S. (1978) Leviathan, open society, and the crisis of ecology, *Western Political Quarterly* 31.

Ostrom, E. (1990) *Governing the Commons: The Evolution of Institutions for Collective Action*, Cambridge University Press, Cambridge.

Parnwell, M.J.G. and King, V.T. (1995) Environmental degradation, resource scarcity and population movement among the Iban of Sarawak, paper presented at the European Association of South-East Asian Studies conference, Leiden, The Netherlands, July.

Pathak, A. (1994) *Contested Domains: The State, Peasants and Forests in Contemporary India*, Sage, London.

Pearce, D.W., Markandya, A. and Barbier, E.B. (1989) *Blueprint for a Green Economy*, Earthscan, London.

Pearson, C.S. (ed.) (1987) *Multinational Corporations, Environment, and the Third World: Business Matters*, Duke University Press, Durham, North Carolina.

Peet, R. (1991) *Global Capitalism: Theories of Societal Development*, Routledge, London.

Peet, R. and Watts, M. (1993) Introduction: development theory and environment in an age of market triumphalism, *Economic Geography* 69, 227–53.

—— (eds) (1996a) *Liberation Ecologies: Environment, Development, Social Movements*, Routledge, London.

—— (1996b) Liberation ecology: development, sustainability, and environment in an age of market triumphalism, in R. Peet and M. Watts (eds), *Liberation Ecologies: Environment, Development, Social Movements*, Routledge, London, pp. 1–45.

Peluso, N.L. (1992) *Rich Forests, Poor People: Resource Control and Resistance in Java*, University of California Press, Berkeley.

—— (1993a) The political ecology of extraction and extractive reserves in East Kalimantan, Indonesia, *Development and Change* 23, 49–74.

—— (1993b) Coercing conservation? The politics of state resource control, *Global Environmental Change* 3, 199–217.

—— (1995) Whose woods are these? Counter-mapping forest territories in Kalimantan, Indonesia, *Antipode* 27, 383–406.

Pepper, D. (1993) *Eco-socialism: From Deep Ecology to Social Justice*, Routledge, London.

Permpongsacharoen, W. (1992) Alternatives from the Thai environmental movement, *Nature and Resources* 28 (2), 4–13.

Peters, P.E. (1984) Struggles over water, struggles over meaning: cattle, water and the state in Botswana, *Africa* 54, 29–49.

—— (1994) *Dividing the Commons: Politics, Policy and Culture in Botswana*, University Press of Virginia.

Pickering, K.T. and Owen, L.A. (1994) *An Introduction to Global Environmental Issues*, Routledge, London.

Piddington, K. (1992) The role of the World Bank, in A. Hurrell and B. Kingsbury (eds), *The International Politics of the Environment: Actors, Interests, and Institutions*, Clarendon Press, Oxford, pp. 212–27.

Porter, G. and Brown, J.W. (1991) *Global Environmental Politics*, Westview Press, Boulder, Colorado.

Pretty, J.N. (1995) *Regenerating Agriculture: Policies and Practice for Sustainability and Self-reliance*, Earthscan, London.

Price, M. (1994) Ecopolitics and environmental non-governmental organisations in Latin America, *Geographical Review* 84, 42–58.

Princen, T. (1994a) NGOs: creating a niche in environmental diplomacy, in T. Princen and M. Finger (eds), *Environmental NGOs in World Politics: Linking the Local and the Global*, Routledge, London, pp. 29–47.

—— (1994b) The ivory ban trade: NGOs and international conservation, in T. Princen and M. Finger (eds), *Environmental NGOs in World Politics: Linking the Local and the Global*, Routledge, London, pp. 121–59.

Princen, T. and Finger, M. (eds) (1994) *Environmental NGOs in World Politics: Linking the Local and the Global*, Routledge, London.

Puntasen, Apichai, Siriprachai, S. and Puyasavatsut, C. (1992) Political economy of eucalyptus: business, bureaucracy and the Thai government, *Journal of Contemporary Asia* 22, 187–206.

Putz, F.E. and Holbrook, N.M. (1988) Tropical rain forest images, in J.S. Denslow and C. Padoch (eds), *People of the Tropical Rain Forest*, University of California Press, Berkeley, pp. 37–52.

Rangan, H. (1996) From Chipko to Uttaranchal: development, environment, and social protest in the Garhwal Himalayas, India, in R. Peet and M. Watts (eds), *Liberation Ecologies: Environment, Development, Social Movements*, Routledge, London, pp. 205–26.

Rangarajan, M. (1996) *Fencing the Forest: Conservation and Ecological Change in India's Central Provinces, 1860–1914*, Oxford University Press, Delhi.

Redclift, M. (1984) *Development and the Environmental Crisis: Red or Green Alternatives?*, Methuen, London.

—— (1987) *Sustainable Development: Exploring the Contradictions*, Methuen, London.

—— (1992) Sustainable development and popular participation: a framework for analysis, in D. Ghai and J.M. Vivian (eds), *Grassroots Environmental Action: People's Participation in Sustainable Development*, Routledge, London, pp. 23–49.

Redclift, M. and Benton, T. (eds) (1994) *Social Theory and the Global Environment*, Routledge, London.

Reed, D. (ed.) (1992) *Structural Adjustment and the Environment*, Earthscan, London.

Rees, W. and Wackernagel, M. (1994) Ecological footprints and appropriated carrying capacity: measuring the natural capital requirements of the human economy, in A. Jansson, M. Hammer, C. Folke and R. Costanza (eds), *Investing in Natural Capital: The Ecological Economics Approach to Sustainability*, Island Press, Washington, DC.

Repetto, R. (1988) Overview, in R. Repetto and M. Gillis (eds), *Public Policies and the Misuse of Forest Resources*, Cambridge University Press, Cambridge, pp. 1–41.

Repetto, R. and Gillis, M. (eds) (1988) *Public Policies and the Misuse of Forest Resources*, Cambridge University Press, Cambridge.

Ribot, J.C. (1995) From exclusion to participation: turning Senegal's forestry policy around?, *World Development* 23, 1587–99.

Rich, B. (1994) *Mortgaging the Earth: The World Bank, Environmental Impoverishment and the Crisis of Development*, Earthscan, London.

Richards, P. (1985) *Indigenous Agricultural Revolution: Ecology and Food Production in West Africa*, Hutchinson, London.

Rigg, J. (1991) Thailand's Nam Choan dam project: a case study in the 'greening' of South-East Asia, *Global Ecology and Biogeography Letters* 1, 42–54.

—— (1993) Forests and farmers, lands and livelihoods: changing resource realities in Thailand, *Global Ecology and Biogeography Letters* 3, 277–89.

—— (1997) *Chasing the Wind: Modernization and Development in South-East Asia*, Routledge, London.

Rigg, J. and Stott, P. (1996) Forest tales: politics, environmental policies and their implementation in Thailand, in U. Desai (ed.), *Comparative Environmental Policy and Politics*, SUNY, New York.

Rocheleau, D. and Ross, L. (1995) Trees as tools, trees as text: struggles over resources in Zambrana-Chacuey, Dominican Republic, *Antipode* 27, 407–28.

Rocheleau, D.E., Steinberg, P.E. and Benjamin, P.A. (1995) Environment, development, crisis, and crusade: Ukambani, Kenya, 1890–1990, *World Development* 23, 1037–51.

Rocheleau, D., Thomas-Slayter, B. and Wangari, E. (eds) (1996) *Feminist Political Ecology: Global Issues and Local Experience*, Routledge, London.

Roussopoulos, D.I. (1993) *Political Ecology: Beyond Environmentalism*, Black Rose, London.

Rowell, A. (1995) Oil, Shell and Nigeria, *The Ecologist* 25, 210–13.

Rush, J. (1991) *The Last Tree: Reclaiming the Environment in Tropical Asia*, Asia Society, New York.

Ryle, M. (1988) *Ecology and Socialism*, Rodins, London.

Sachs, W. (ed.) (1992) *The Development Dictionary: A Guide to Knowledge as Power*, Zed Books, London.

—— (ed.) (1993) *Global Ecology: A New Arena of Political Conflict*, Zed Books, London.

Said, E.W. (1978) *Orientalism*, Penguin, Harmondsworth.

Saldana, I.M. (1990) The political ecology of traditional farming practices in Thana District, Maharastra (India), *Journal of Peasant Studies* 17, 433–43.

Sanyal, B. (1994) *Cooperative Autonomy: The Dialectic of State–NGOs relationship in Developing Countries*, International Institute for Labour Studies, Geneva.

217

Sargent, C. and Bass, S. (eds) (1992) *Plantation Politics: Forest Plantations in Development*, Earthscan, London.

Schmidheiny, S. (1992) *Changing Course: A Global Perspective on Development and the Environment*, MIT Press, Cambridge, Massachusetts.

Schmink, M. (1988) Big business in the Amazon, in J.S. Denslow and C. Padoch (eds), *People of the Tropical Rain Forest*, University of California Press, Berkeley, pp. 163–76.

Schmink, M. and Wood, C.H. (1987) The 'political ecology' of Amazonia, in P.D. Little and M.M. Horowitz (eds), *Lands at Risk in the Third World: Local-level Perspectives*, Westview Press, Boulder, Colorado, pp. 38–57.

—— (1992) *Contested Frontiers in Amazonia*, Columbia University Press, New York.

Schroeder, R.A. (1993) Shady practice: gender and the political ecology of resource stabilization in Gambian garden/orchards, *Economic Geography* 69, 349–65.

—— (1995) Contradictions along the commodity road to environmental stabilization: foresting Gambian gardens, *Antipode* 27, 325–42.

Scott, A. (1990) *Ideology and the New Social Movements*, Unwin Hyman, London.

Scott, J.C. (1976) *The Moral Economy of the Peasant: Rebellion and Subsistence in Southeast Asia*, Yale University Press, New Haven.

—— (1985) *Weapons of the Weak: Everyday Forms of Peasant Resistance*, Yale University Press, New Haven.

—— (1990) *Domination and the Arts of Resistance: Hidden Transcripts*, Yale University Press, New Haven.

Secrett, C. (1987) Friends of the Earth UK and the hardwood campaign, in Sahabat Alam Malaysia (SAM) (ed.), *Proceedings of the Conference on Forest Resources Crisis in the Third World 6–8 September 1986*, SAM, Penang, pp. 348–56.

Sen, G. and Grown, C. (1987) *Development, Crises and Alternative Visions: Third World Women's Perspectives*, Earthscan, London.

Sesmou, K. (1991) The Food and Agriculture Organization of the United Nations: an insider's view, *The Ecologist* 21, 47–56.

Setchell, C.A. (1995) The growing environmental crisis in the world's mega-cities: the case of Bangkok, *Third World Planning Review* 17, 1–18.

Sheridan, T.E. (1988) *Where the Dove Calls: The Political Ecology of a Peasant Corporate Community in Northwestern Mexico*, University of Arizona Press, Tucson.

Shiva, V. (1988) *Staying Alive: Women, Ecology and Development*, Zed Books, London.

—— (1991a) *Ecology and the Politics of Survival: Conflicts over Natural Resources in India*, Sage, London.

—— (1991b) The Green Revolution in the Punjab, *The Ecologist* 21, 57.

—— (1992) The road from Rio: 'greenwash' at the Earth Summit, *Frontline* (July 3), 105–8.

—— (1993) The greening of the global reach, in W. Sachs (ed.), *Global Ecology: A New Arena of Political Conflict*, Zed Books, London, pp. 149–56.

Silva, E. (1994) Thinking politically about sustainable development in the tropical forests of Latin America, *Development and Change* 25, 697–721.

Simmonds, I.G. (1993) *Interpreting Nature: Cultural Constructions of the Environment*, Routledge, London.

Sinaga, K. (1994) *NGOs in Indonesia: A Study of the Role of Non-governmental Organizations in the Development Process*, Verlag für Entwicklungspolitik Breitenbach Gmbh, Saarbrucken, Germany.

Skocpol, T. (1985) Bringing the state back in: strategies of analysis in current research, in P.B. Evans, D. Rueschemeyer and T. Skocpol (eds), *Bringing the State Back In*, Cambridge University Press, Cambridge, pp. 3–37.

Smil, V. (1984) *The Bad Earth: Environmental Degradation in China*, Zed Books, London.

Smillie, I. (1995) *The Alms Bazaar: Altruism under Fire – Non-profit Organizations and International Development*, IT Publications, London.

Smith, M. (1994) *Paradise Lost? The Suppression of Environmental Rights and Freedom of Expression in Burma*, Article 19, London.

Sontheimer, S. (ed.) (1991) *Women and the Environment: A Reader*, Earthscan, London.

Stauber, J.C. and Rampton, S. (1995) 'Democracy' for hire: public relations and environmental movements, *The Ecologist* 25, 173–80.

Steiner, D. and Nauser, M. (eds) (1993) *Human Ecology: Fragments of Anti-Fragmentary Views of the World*, Routledge, London.

Stoett, P.J. (1993) International politics and the protection of great whales, *Environmental Politics* 2, 277–303.

Stonich, S. (1993) *'I am Destroying the Land': The Political Ecology of Poverty and Environmental Destruction in Honduras*, Westview Press, Boulder, Colorado.

Susman, P., O'Keefe, P. and Wisner, B. (1983) Global disasters: a radical interpretation, in K. Hewitt (ed.), *Interpretations of Calamity from the Viewpoint of Human Ecology*, Allen and Unwin, London, pp. 263–83.

Swift, A. (1993) *Global Political Ecology: The Crisis in Economy and Government*, Pluto Press, London.

Swyngedouw, E.A. (1995) The contradictions of urban water provision: a study of Guayaquil, Ecuador, *Third World Planning Review* 17, 387–406.

—— (1996) Power, nature and the city: the conquest of water and the political-ecology of urbanization in Guayaquil, Ecuador, *Environment and Planning A*.

Szerszynski, B., Lash, S. and Wynne, B. (1996) Introduction: ecology, realism and the social sciences, in S. Lash, B. Szerszynski and B. Wynne (eds), *Risk, Environment and Modernity: Towards a New Ecology*, Sage, London, pp. 1–26.

Taylor, J.G. (1979) *From Modernization to Modes of Production: A Critique of the Sociologies of Development and Underdevelopment*, Macmillan, London.

Taylor, R.H. (1987) *The State in Burma*, C. Hurst, London.

Thirgood, J.V. (1981) *Man and the Mediterranean Forest: A History of Resource Depletion*, Academic Press, London.

Thrupp, L.A. (1990) Environmental initiatives in Costa Rica: a political ecology perspective, *Society and Natural Resources* 3, 243–56.

Tickell, O. and Hildyard, N. (1992) Green dollars, green menace, *The Ecologist* 22, 82–3.

Toure, O. (1988) The pastoral environment of northern Senegal, *Review of African Political Economy* 42, 32–9.

Toye, J. (1993) *Dilemmas of Development: Reflections on the Counter-revolution in Development Economics*, second edition, Blackwell, Oxford.

Turner, B.L. *et al.* (1990) Two types of global environmental change: definitional and spatial-scale issues in their human dimensions, *Global Environmental Change* 1, 14–22.

Turner, M. (1993) Overstocking the range: a critical analysis of the environmental science of Sahelian pastoralism, *Economic Geography* 69, 402–21.

Turner, R.K. (1995) Environmental economics and management, in T. O'Riordan (ed.), *Environmental Science for Environmental Management*, Longman, London, pp. 30–44.

Usher, A.D. (1996) The race for power in Laos: the Nordic connection, in M.J.G. Parnwell and R.L. Bryant (eds), *Environmental Change in South-East Asia: People, Politics and Sustainable Development*, Routledge, London, pp. 123–44.

Utting, P. (1993) *Trees, People and Power: Social Dimensions of Deforestation and Forest Protection in Central America*, Earthscan, London.

Vandergeest, P. and Peluso, N.L. (1995) Territorialization and state power in Thailand, *Theory and Society* 24, 385–426.

Vayda, A.P. (1983) Progressive contextualization: methods for research in human ecology, *Human Ecology* 11, 265–81.

Vitug, M.D. (1993) *The Politics of Logging: Power from the Forest*, Philippine Center for Investigative Journalism, Manila.

Vivian, J.M. (1994) NGOs and sustainable development in Zimbabwe: No magic bullets, *Development and Change* 25, 167–93.

Vogler, J. (1995) *The Global Commons: A Regime Analysis*, Wiley, Chichester.

Vogler, J. and Imber, M.F. (eds) (1996) *The Environment and International Relations*, Routledge, London.

Walker, K.J. (1988) The environmental crisis: a critique of neo-Hobbesian responses, *Polity* 21, 67–81.

—— (1989) The state in environmental management: the ecological dimension, *Political Studies* 37, 25–38.

Walker, R.B.J. (1993) *Inside/Outside: International Relations as Political Theory*, Cambridge University Press, Cambridge.

Wallerstein, I. (1974) *The Modern World System, Capitalist Agriculture and the Origins of the European World Economy in the Sixteenth Century*, Academic Press, New York.

Walpole, P., Braganza, G., Ong, J.B., Tengco, G.J. and Wijanco, E. (1993) *Upland Philippine Communities: Guardians of the Final Forest Frontiers*, Berkeley: Center for Southeast Asia Studies Research Network Report no. 4, University of California.

Wapner, P. (1995) Politics beyond the state: environmental activism and world civic politics, *World Politics* 47, 311–40.

—— (1996) *Environmental Activism and World Civic Politics*, State University of New York Press, Albany.

Waterbury, J. (1979) *Hydropolitics of the Nile Valley*, Syracuse University Press, Syracuse, New York.

Watts, M. (1983a) *Silent Violence: Food, Famine and Peasantry in Northern Nigeria*, University of California Press, Berkeley.

—— (1983b) On the poverty of theory: natural hazards research in context, in K. Hewitt (ed.), *Interpretations of Calamity from the Viewpoint of Human Ecology*, Allen and Unwin, London, pp. 231–62.

—— (1984) The demise of the moral economy: food and famine in a Sudano-Sahelian region in historical perspective, in E. Scott (ed.), *Life Before the Drought*, Allen and Unwin, London, pp. 124–48.

—— (1989) The agrarian question in Africa: debating the crisis, *Progress in Human Geography* 13, 1–41.

—— (1993) Development I: power, knowledge, discursive practice, *Progress in Human Geography* 17, 257–72.

—— (1994) Development II: the privatization of everything?, *Progress in Human Geography* 18, 371–84.

Watts, M. and Bohle, H.G. (1993) The space of vulnerability: the causal structure of hunger and famine, *Progress in Human Geography* 17, 43–67.

Watts, M. and Peet, R. (eds) (1993) Environment and development, parts I and II, Special issue, *Economic Geography* 69 (3/4), 227–421.

—— (1996) Conclusion: towards a theory of liberation ecology, in R. Peet and M. Watts (eds), *Liberation Ecologies: Environment, Development, Social Movements*, Routledge, London, pp. 260–9.

Weale, A. (1992) *The New Politics of Pollution*, Manchester University Press, Manchester.

Weinberg, B. (1991) *War on the Land: Ecology and Politics in Central America*, Zed Books, London.

Weir, D. (1988) *The Bhopal Syndrome*, Earthscan, London.

Welford, R. (ed.) (1996) *Corporate Environmental Management: Systems and Strategies*, Earthscan, London.

Westing, A.H. (1992) Environmental refugees: a growing category of displaced persons, *Environmental Conservation* 19, 201–7.

Westoby, J. (1987) *The Purpose of Forests: Follies of Development*, Basil Blackwell, Oxford.

Weston, J. (ed.) (1986) *Red or Green*, Pluto Press, London.

White, D. (1994) The environment in focus, *RTZ Review* 31, 3–5.

Whyte, A.V.T. (1986) From hazard perception to human ecology, in R.W. Kates and I. Burton (eds), *Geography, Resources, and Environment*, vol. 2, University of Chicago Press, Chicago, pp. 240–71.

Williams, M. (1993) *International Economic Organisations and the Third World*, Harvester Wheatsheaf, Hemel Hempstead.

Wilson, G.A. and Bryant, R.L. (1997) *Environmental Management: New Directions for the 21st Century*, UCL Press, London.

Wisner, B. (1976) *Man-made Famine in Eastern Kenya: The Interrelationship of Environment and Development*, IDS discussion paper no. 96, IDS, Brighton.

—— (1978) Does radical geography lack an approach to environmental relations?, *Antipode* 10, 84–95.

Wisner, B., Weiner, D. and O'Keefe, P. (1982) Hunger: a polemical review, *Antipode* 14, 1–16.

Wolf, E. (1972) Ownership and political ecology, *Anthropological Quarterly* 45, 201–5.

—— (1982) *Europe and the People without History*, University of California Press, Berkeley.

World Bank (1992) *World Development Report 1992: Development and the Environment*, Oxford University Press, Oxford.

World Commission on Environment and Development (1987) *Our Common Future*, Oxford University Press, Oxford.

World Rainforest Movement [WRM] and Sahabat Alam Malaysia [SAM] (1989) *The Battle for Sarawak's Forests*, WRM and SAM, Penang.

World Resources Institute (1990) *World Resources 1990–91*, Oxford University Press, Oxford.

Worster, D. (1985) *Rivers of Empire: Water, Aridity, and the Growth of the American West*, Pantheon, New York.

—— (1988) Doing environmental history, in D. Worster (ed.), *The Ends of the Earth: Perspectives on Modern Environmental History*, Cambridge University Press, Cambridge, pp. 289–307.

Yabes, R.A. (1992) The *zanjeras* and the Ilocos Norte irrigation project: lessons of environmental sustainability from Philippine traditional resource management systems, in D. Ghai and J.M. Vivian (eds), *Grassroots Environmental Action: People's Participation in Sustainable Development*, Routledge, London, pp. 106–40.

Yap, N.T. (1989/90) NGOs and sustainable development, *International Journal* 45, 75–105.

Yapa, L. (1979) Ecopolitical economy of the green revolution, *The Professional Geographer* 31, 371–6.

You, Jong-Il (1995) The Korean model of development and its environmental implications, in V. Bhaskar and A. Glyn (eds), *The North, the South, and the Environment: Ecological Constraints and the Global Economy*, Earthscan, London, pp. 158–83.

Young, O.R. (1989) *International Cooperation: Building Regimes for Natural Resources and the Environment*, Cornell University Press, Ithaca, New York.

Young, S.C. (1992) The different dimensions of green politics, *Environmental Politics* 1, 9–44.

Zimmerer, K.S. (1993) Soil erosion and social (dis)courses in Cochabamba, Bolivia: perceiving the nature of environmental degradation, *Economic Geography* 69, 312–27.

—— (1994) Human geography and the 'new ecology': the prospect and promise of integration, *Annals of the Association of American Geographers* 84, 108–25.

INDEX

223

Printed in the United States
133906LV00002B/34-42/A